Irmgard Riedmaier

**Development of mRNA patterns for screening of anabolic steroids**

Irmgard Riedmaier

# Development of mRNA patterns for screening of anabolic steroids

Südwestdeutscher Verlag für Hochschulschriften

**Impressum/Imprint (nur für Deutschland/ only for Germany)**
Bibliografische Information der Deutschen Nationalbibliothek: Die Deutsche Nationalbibliothek verzeichnet diese Publikation in der Deutschen Nationalbibliografie; detaillierte bibliografische Daten sind im Internet über http://dnb.d-nb.de abrufbar.

Alle in diesem Buch genannten Marken und Produktnamen unterliegen warenzeichen-, marken- oder patentrechtlichem Schutz bzw. sind Warenzeichen oder eingetragene Warenzeichen der jeweiligen Inhaber. Die Wiedergabe von Marken, Produktnamen, Gebrauchsnamen, Handelsnamen, Warenbezeichnungen u.s.w. in diesem Werk berechtigt auch ohne besondere Kennzeichnung nicht zu der Annahme, dass solche Namen im Sinne der Warenzeichen- und Markenschutzgesetzgebung als frei zu betrachten wären und daher von jedermann benutzt werden dürften.

Verlag: Südwestdeutscher Verlag für Hochschulschriften Aktiengesellschaft & Co. KG
Dudweiler Landstr. 99, 66123 Saarbrücken, Deutschland
Telefon +49 681 37 20 271-1, Telefax +49 681 37 20 271-0
Email: info@svh-verlag.de
Zugl.: Freising, TUM, Diss., 2009

Herstellung in Deutschland:
Schaltungsdienst Lange o.H.G., Berlin
Books on Demand GmbH, Norderstedt
Reha GmbH, Saarbrücken
Amazon Distribution GmbH, Leipzig
ISBN: 978-3-8381-1300-5


**Imprint (only for USA, GB)**
Bibliographic information published by the Deutsche Nationalbibliothek: The Deutsche Nationalbibliothek lists this publication in the Deutsche Nationalbibliografie; detailed bibliographic data are available in the Internet at http://dnb.d-nb.de.

Any brand names and product names mentioned in this book are subject to trademark, brand or patent protection and are trademarks or registered trademarks of their respective holders. The use of brand names, product names, common names, trade names, product descriptions etc. even without a particular marking in this works is in no way to be construed to mean that such names may be regarded as unrestricted in respect of trademark and brand protection legislation and could thus be used by anyone.

Publisher: Südwestdeutscher Verlag für Hochschulschriften Aktiengesellschaft & Co. KG
Dudweiler Landstr. 99, 66123 Saarbrücken, Germany
Phone +49 681 37 20 271-1, Fax +49 681 37 20 271-0
Email: info@svh-verlag.de

Printed in the U.S.A.
Printed in the U.K. by (see last page)
ISBN: 978-3-8381-1300-5


Copyright © 2010 by the author and Südwestdeutscher Verlag für Hochschulschriften Aktiengesellschaft & Co. KG and licensors
All rights reserved. Saarbrücken 2010

# Table of Contents

Abbreviations ............................................................................................... 3
Zusammenfassung ....................................................................................... 5
Abstract ....................................................................................................... 7

**1 Introduction** ........................................................................................ 9
   1.1 Anabolic Steroid Hormones – use and misuse in animal husbandry .... 9
   1.2 Steroid Hormones in Hormone Replacement Therapy ..................... 10
   1.3 Potential of transcriptomics for biomarker development to trace
        anabolic steroid hormone functions .................................................. 10
   1.4 Aims ................................................................................................... 12

**2 Materials and Methods** ..................................................................... 14
   2.1 Animal Experiments ........................................................................... 14
   2.2 RNA Extraction and Quality Determination ....................................... 17
   2.3 Selection of Target Genes ................................................................. 19
   2.4 Specific Primer Design ...................................................................... 23
   2.5 Two-Step RT-qPCR Analysis ............................................................. 23
   2.6 One-Step RT-qPCR Analysis ............................................................. 24
   2.7 Data Analysis and Statistics .............................................................. 25

**3 Results and Discussion** ................................................................... 28
   3.1 Anabolics study on Nguni Cattle ....................................................... 28
   3.2 Pour on anabolics study in veal calves ............................................. 37
   3.3 SARM Study on *Macaca fascicularis* ............................................... 41

**4 Conclusions and Perspectives** ....................................................... 46

**5 References** ........................................................................................ 50

Acknowledgements .................................................................................... 59
Scientific Communication .......................................................................... 60
Appendix .................................................................................................... 62

# Abbreviations

| | | | |
|---|---|---|---|
| aCP1 | acid phosphatase 1 | GR | glucocorticoid receptor |
| ACTA2 | actinα 1 | HRE | hormone responsive element |
| ACTB | actinβ | IFN | interferone |
| ADRBK2 | adrenergic β kinase 2 | IGF-1 | insulin like growth factor 1 |
| AR | androgen receptor | IGF-1R | insulin like growth factor 1 receptor |
| BCL-2 | B-cell CLL/lymphoma 2 | | |
| BCL-XL | B-cell lymphoma extra large | IGFBP3 | insulin like growth factor binding protein 3 |
| bp | base pairs | | |
| C | control | IL | interleukin |
| Casp | caspase | LAP | lingual antimicrobial peptide |
| CC | carrier control | LTF | lactoferrin |
| cDNA | complementary DNA | MGA | melengestrolacetate |
| CK | creatine kinase | MHC | major histocompatibility complex |
| CK8, 18 | cytokeratin kinase 8, 18 | mRNA | messenger RNA |
| CP2 | transcription factor CP2 | MTPN | myotropin |
| Ct | threshold cycle | NTC | no template control |
| DEGMBE | diethylenglycolmonobutylether | OD | optical density |
| DMSO | dimethylsulfoxid | PCA | principal components analysis |
| DNA | desoxyribonucleic acid | PCR | polymerase chain reaction |
| dNTP | desoxyribonucleosidtriphosphate | pmol | picomol |
| EGF | epidermal growth factor | PR | progesterone receptor |
| EGFR | epidermal growth factor receptor | RT-qPCR | quantitative reverse transcription-polymerase chain reaction |
| EITR | estrogen induced transcription factor | | |
| | | RBM | RNA binding protein |
| ER | estrogen receptor | rev | reverse |
| Fas | TNF receptor superfamily member 6 | RG | reference gene |
| | | RIN | RNA integrity number |
| FasL | TNF receptor superfamily member 6 ligand | RNA | ribonucleic acid |
| | | RSA | Republic of South Africa |
| FGF | fibrobast growth factor | RT | reverse transcription |
| FGFBP | fibrobast growth factor binding protein | SARM | selective androgen receptor modulator |
| for | forward | SD | standard deviation |
| GAPDH | glycerinealdehyde-3-phosphate dehydrogenase | SERM | selective estrogen receptor modulator |

| | |
|---|---|
| T1 | one time treated group |
| T3 | three times treated group |
| TBA | trenbolone acetate |
| Testo | testosterone |
| TGF | tumor growth factor |
| TMOD | tropomodulin |
| TNF | tumor necrose factor |
| TNFR | tumor necrose factor receptor |
| UB3 | ubiquitin 3 |
| USF | upstream transcription factor |
| YWHAZ | tyrosine 3-monooxygenase/tryptophan 5-monooxygenase activation protein, $\zeta$ polypeptide |

# Zusammenfassung

Natürliche Steroidhormone werden ausgehend von Cholesterin gebildet und sind in die endo- und parakrine Wachstumsregulation verschiedener Gewebe involviert. Einzelne Steroidhormone, wie Östrogene und Androgene wirken anabol, indem sie die Proteinretention im Körper verbessern und Fettreserven abbauen, was zu einer Erhöhung der Wachstumsrate führt. In der Tiermast werden diese anabolen Eigenschaften genutzt, um die Gewichtszunahme und die Futterverwertung zu verbessern, womit die Produktivität erhöht und Kosten gesenkt werden.

In einigen Ländern, wie den USA, Kanada, Australien, Mexiko und Südafrika ist der Gebrauch von Wachstumsförderern in der Tiermast zugelassen. Aufgrund erwiesener Nebenwirkungen für den Konsumenten ist der Gebrauch anaboler Substanzen in der EU verboten, wo die Einhaltung dieser Richtlinie (88/146/EEC) streng überwacht wird.

Ein weiteres Anwendungsgebiet anaboler Steroidhormone ist die Behandlung altersbedingter Krankheiten, wie Osteoporose oder Sarkopenie, welche durch eine Abnahme der endogenen Produktion von Östradiol und Testosteron bei rückläufiger Gonadenfunktion verursacht werden. Für die Behandlung dieser altersbedingten Krankheiten wurden so genannte selektive Androgen Rezeptor Modulatoren (SARM) entwickelt. Darunter versteht man synthetische Moleküle, welche die nützlichen zentralen und peripheren Eigenschaften von Testosteron besitzen, jedoch kaum Nebenwirkungen aufweisen. Aufgrund der positiven Wirkungen eines SARM auf die Muskelmasse ist das Risiko des Missbrauchs dieser Substanzen in der Tiermast oder im Sport vorhanden.

Um den Missbrauch anaboler Substanzen in der Tiermast oder im Sport zu kontrollieren, werden Hormonrückstände mittels Immunoassays oder chromatographischer Methoden in Kombination mit Massenspektrometrie detektiert. Mit Hilfe dieser Methoden können nur bekannte Substanzen nachgewiesen werden. Um eine effiziente Kontrolle des Missbrauchs anaboler Stoffe zu gewährleisten, ist es nötig neue Technologien zu entwickeln, mit welchen man den Gebrauch einer breiten Masse an illegalen Medikamenten, inklusive neu entwickelter Xenobiotika nachweisen kann.

Ein Ansatz zur Entwicklung einer neuen Nachweismethode ist das Aufzeigen physiologischer Effekte, welche durch die Einnahme anaboler Substanzen verursacht werden. Ein viel versprechender Weg solche physiologischen Veränderungen nachzuweisen, ist die Bestimmung von Veränderungen in der mRNA Expression mittels quantitativer real-time RT-PCR (RT-qPCR).

Ziel dieser Arbeit war es zu prüfen, ob die Bestimmung von Genexpressionsveränderungen Potential für die Entwicklung neuer Technologien zum Nachweis missbräuchlicher Anwendung anaboler Substanzen hat. Hierfür wurde die mRNA Expression steroidabhängiger Gene im Blut und in vaginalen Epithelzellen – Gewebe, welche leicht vom lebenden Individuum genommen werden können - mittels RT-qPCR quantifiziert und mögliche Veränderungen statistisch bewertet.

In allen drei Tierversuchen, die im Rahmen dieser Arbeit durchgeführt wurden, konnten Genexpressionsveränderungen festgestellt werden. In zwei dieser Studien konnte aus den signifikant regulierten Genen mit Hilfe biostatistischer Methoden, wie der Principal Components Analyse (PCA) oder der hierarchischen Clusteranalyse eine Trennung von Kontrollgruppe und Behandlungsgruppe dargestellt werden.

Die Ergebnisse dieser Arbeit zeigen, dass die Quantifizierung von Genexpressionsveränderungen eine vielversprechende Herangehensweise für die Entwicklung neuer Technologien zum Nachweis des missbräuchlichen Gebrauchs anaboler Substanzen darstellt.

# Abstract

Natural steroid hormones are synthesized from cholesterol and are involved in endocrine and paracrine regulation of growth in different tissues. Some steroid hormones like androgens and estrogens have anabolic functions by enhancing body protein accretion and mobilizing fat stores, which results in an increased growth rate. These properties are useful in animal husbandry to improve weight gain and feed efficiency and thereby increase productivity and reduce costs. In some countries like the USA, Canada, Australia, Mexico and South Africa the use of growth promoters is approved. Due to proven negative side effects for consumers the use of anabolic substances is forbidden in the EU, where the compliance of this directive (88/146/EEC) is strictly controlled.

Another application area of anabolic steroid hormones is the treatment of age related diseases like osteoporosis or sarcopenia, which are related to a decrease of the endogenous production of anabolic steroid hormones during diminishing gonade function, mainly estradiol and testosterone. For the treatment of these age related diseases, synthetic molecules called selective androgen receptor modulators (SARMs) are developed, which have the potential to mimic the desirable central and peripheral androgenic anabolic effects of testosterone but with less side effects. Due to the positive effects on muscle strength of SARMs the risk of the misuse of these substances in animal husbandry or human sports as anabolic agent is present.

To uncover the abuse of anabolic agents in animal husbandry or human sports, hormone residues are detected by immuno assays or chromatographical methods in combination with mass spectrometry. With these methods only known substances can be discovered. To enable an efficient tracing of unknown misused anabolic substances it is necessary to develop new technologies to screen for a broad range of illegal drugs including newly designed xenobiotic anabolic agents.

Verifying physiological effects caused by anabolic agents will be a new way to develop potential monitoring systems. The determination of changes in mRNA expression by quantitative real-time RT-PCR (RT-qPCR) is a promising approach to verify those physiological changes.

The aim of this thesis was to proof the potential of the determination of mRNA expression changes for the development of a screening method to detect the misuse of anabolic steroid hormones. Therefore expression changes of steroid responsive genes that were selected by screening the actual literature were quantified in blood and vaginal epithelial cells – tissues that can easily been taken from the living individual. Gene expression changes were measured by RT-qPCR.

In all three animal trials included in this thesis, expression changes of multiple genes in blood and bovine vaginal smear could be quantified. In two studies, biostatistical tools, like Principal Components Analysis (PCA) or Hierarchical Cluster Analysis were successfully used to distinguish treated and untreated animals by involving all significantly regulated genes.

The results of this thesis indicate that the quantification of gene expression changes is a promising approach for the development of new screening methods to trace the abuse of anabolic agents.

# 1 Introduction

## 1.1 Anabolic Steroid Hormones – use and misuse in animal husbandry

Natural steroid hormones are synthesized from cholesterol and can be classified in five subgroups: mineralocorticoids, glucocorticoids, gestagens, androgens, and estrogens and are involved in endocrine and paracrine regulation of different tissues. Some steroid hormones like androgens and estrogens have anabolic functions by enhancing body protein accretion and mobilizing fat stores, which results in an increased growth rate [1, 2]. These properties are deep-rooted in the evolution of vertebrates. The sex steroids testosterone and estradiol have effects on behavioral, morphological and physiological traits. Estrogens stimulate protein- and mineral retention during pregnancy, which is important for the development of the embryo. Testosterone promotes sexual behaviours like courtship and improves growth of skeletal muscle which is important for defending the territory [3, 4].

In animal husbandry the myotropic, growth promoting properties of steroid hormones are beneficial. Used orally, the natural steroid hormones testosterone and estradiol are almost inactive. Besides these natural steroids also the xenobiotic hormones trenbolone acetate (TBA), zeranol and melengestrol acetate (MGA) were developed by US companies to be used as anabolics in food producing animals. As only MGA is orally active, the other drugs have to be applied by implantation [5]. In meat production growth promoters are used to increase productivity and to reduce costs by improving weight gain and feed efficiency [6, 7]. The use of growth promoters is approved in some countries like the USA, Canada, Australia, Mexico, and South Africa. It has been proven that hormone residues in meat are increased and have adverse side effects for the consumer [8-11]. Therefore the use of anabolic agents in meat producing animals and also the import of meat derived from cattle given these substances is forbidden in the EU since 1988. To enforce the directive (88/146/EEC), permanent surveillance is essential [9, 12-16].

## 1.2 Steroid Hormones in Hormone Replacement Therapy

Over the last decades the proportion of elderly people in the population has increased [17]. This is the reason why the incidence of age related conditions like sarcopenia (loss of muscle mass) and osteoporosis (loss of bone density) is rising and becoming one of the major topics in health care [18-21]. The combination of sarcopenia and osteoporosis results in a high incidence of bone fractures relating to accidental falls, which is a significant cause of morbidity in the elderly population. Both conditions are related to the decrease in the endogenous production of anabolic steroid hormones, mainly estradiol and testosterone [22]. Hormone replacement therapy is a major topic in the treatment of frailty. Men and women suffering from frailty are treated with testosterone or estradiol but both therapies are associated with various side effects, like skin virilization in women, prostate hypertrophy in men and an increased risk of cancer [23-25]. An alternative to the treatment with natural testosterone or estradiol are synthetic molecules called SARM (selective androgen receptor modulators) and SERM (selective estrogen receptor modulators), which bind to the steroid hormone receptors exhibiting predominantly tissue selective effects [26]. An "ideal" SARM or SERM is an orally active compound that provides an increase in muscle mass and strength and has a positive effect on bone density without inducing undesirable side effects [27]. Due to the positive effects on muscle strength of SARMs and SERMs the risk of the misuse of these substances as anabolic agent is present.

## 1.3 Potential of transcriptomics for biomarker development to trace anabolic steroid hormone functions

To uncover the abuse of anabolic agents in animal husbandry hormone residues are detected using immuno assays or chromatographical methods in combination with mass spectrometry [28-31]. With these methods only known substances can be discovered. To enable an efficient tracing of misused anabolic substances, it is necessary to develop new technologies to screen for a broad range of illegal drugs including newly designed xenobiotic anabolic agents.

In molecular medicine, e.g in cancer research, the development of molecular biomarkers is already a common approach in diagnostics. Plasma biomarkers are developed for prognostic use and tumor biomarkers are used to develop treatment strategies for each individual patient [32, 33]. To develop such biomarkers *omic* technologies, like transcriptomics, proteomics and metabolomics are applied [34-36].

The use of these *omic* technologies to develop biomarker patterns by tracing anabolic steroid hormone functions will be a promising way to develop a new screening method for the detection of the misuse of anabolic agents [37].

Steroid hormone receptors belong to the family of nuclear receptors and show a high affinity to their corresponding hormone [38, 39]. They are either localized in the cytoplasm moving to the cell nucleus upon activation or directly in the nucleus waiting for the steroid hormones or active analoga to enter the nucleus and activate them [40]. Without a bound ligand the steroid receptors exist as a steroid receptor complex, associated with different heat shock proteins (hsp90, hsp 56, hsp70) and p23 [41-43]. Binding of the ligand results in a conformational change leading to the dissociation of the HSP-complex from the receptor. After dimerization the receptor binds to specific sequences in the promoter region of steroid hormone regulated genes, called hormone responsive elements (HRE) [39, 44, 45]. After DNA binding, different coregulators that are recruited to activate transcription of target genes. Figure 1 shows the main steps in steroid hormone action.

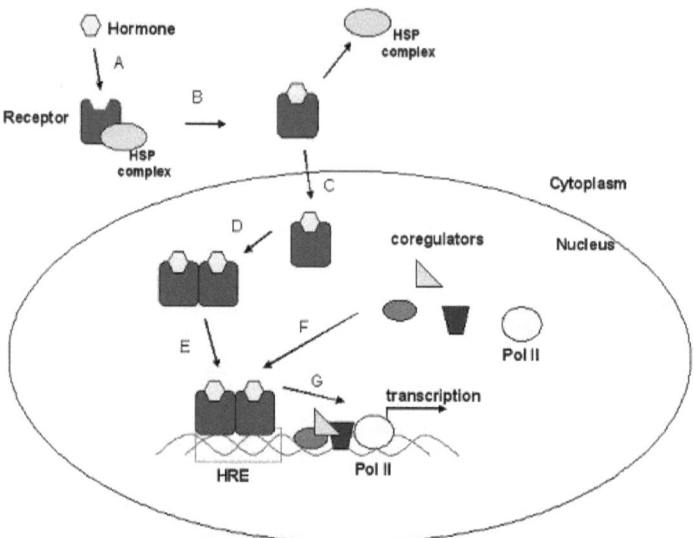

Figure 1: Schematic diagram of the activation of a cytoplasmic steroid hormone receptor
After hormone binding (A) the HSP complex dissociates from the receptor (B), the hormone receptor complex translocates to the nucleus (C), dimerizes (D) and binds to a hormone responsive element (HRE) in the promoter region of a specific gene (E). After binding to the HRE different coregulators of transcription are recruited (F), which are responsible for transcriptional activation [46, 47].

Steroid hormones not solely regulate gene transcription activity, but also influence the stability of generated mRNA. They are able to stabilize or destabilize specific mRNAs. Most is known about the influence of steroid hormones on the stability of their receptor mRNA. Whereas steroid receptor protein is normally down-regulated by their ligands, the regulation of the stability of steroid receptor mRNA may be positive or negative. Regulation of mRNA stability is not restricted to steroid hormone receptors, other genes are also regulated by similar mechanism [48].

## 1.4 Aims

The objective of this thesis was to proof the potential of transcriptomics technology for the development of a screening method to detect the misuse of anabolic steroid hormones. Therefore three different animal trials were employed. Two studies on female cattle, where the effects of different combinations of steroid hormones on gene expression in blood and vaginal smear was quantified and one study on cynomolgus monkeys (*Macaca fascicularis*) where the effects a novel SARM on gene expression of whole blood was compared to the effects of natural testosterone. In all three animal trials expression changes of steroid responsive genes that were selected by screening the actual literature were quantified. Gene expression changes were measured by RT-qPCR. Biostatistical tools, like PCA or Hierarchical Cluster Analysis could be helpful to proof, if quantified gene expression changes will be promising biomarkers for the development of a new screening system to detect the misuse of anabolic agents. Figure 2 presents a schematic overview of the transcriptomic approach to trace anabolic hormone functions.

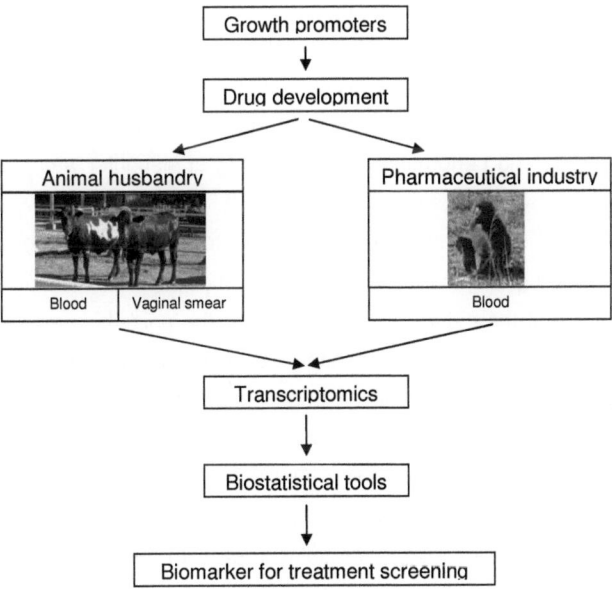

Figure 2: Scheme of the use of transcriptomics to trace anabolic hormone functions.

## 2  Materials and Methods

### 2.1  Animal Experiments

***Anabolics study on Nguni Cattle***
To test the potential to use transcriptomics for the development of a new sensitive test system to screen for the misuse of anabolic agents in food producing animals, an animal study that represents the situation in praxis had to be organized. In the Republic South Africa the use of certain anabolic agents for improvement of growth of meat producing animals is permitted. Anabolic preparations are available on the market there. Therefore it was easier to get the permission for such an animal study in South Africa than it would be in Germany. So it was part of this PhD thesis to organize all steps (study design, sampling of blood and vaginal smear, tissue sampling at slaughter) of an animal study in cooperation with the Onderstepoort Veterinary Institute in Pretoria, South Africa. A journey to Pretoria of two weeks for organizing the transfer of the samples to Germany and for organizing and performing tissue sampling at slaughter was part of this PhD thesis.

Earlier animal studies that were performed to analyze the effects of anabolic substances on growth performance in cattle showed, that 5-10 animals per group are adequate if the natural variance between the animals is minimized by using animals within one breed, one gender and of similar age [49]. It is assumed that gene expression will vary like growth performance and so this aspect was considered in the planning of the study.

Regarding the present conditions in Onderstepoort it was possible to include 18 animals in the trial. The healthy, non pregnant, two year old Nguni heifers were separated into two groups of nine animals each (n = 9). One group was treated with Revalor H® (140 mg Trenbolone acetate plus 14 mg estradiol; Intervet, Isando, RSA) by implantation into the middle third of the pinna of the ear and one group was untreated and served as control.

For this PhD Thesis, whole blood and vaginal smear samples were taken at four time points. Predose samples were taken after study start without prior treatment. Further samples were taken at day 2, day 16 and day 39 of treatment. At the same time points a complete blood count was done by the section of clinical pathology,

University of Pretoria, South Africa, to control the health status of the animals. Blood samples for gene expression analysis were taken as described previously [50]. Vaginal smear was taken using a sterilized spoon. The smear was directly transferred into TriFast (Peqlab, Erlangen, Germany) and stored at -80°C. The animal attendance and blood sampling were done by the Onderstepoort Veterinary Institute (Onderstepoort, Pretoria, South Africa). The animals were housed and fed according to good veterinary practice.

## *Pour on anabolics study on veal calves*

To test the potential of the identified biomarker candidates from the anabolics study on Nguni cattle a second study in which animals are treated with an illicit hormone cocktail was conducted. The application of such hormone cocktail is not allowed, even in countries where the use of anabolic agents in food producing animals is permitted, and so the performance of this trial required a special permission from the government. This trial was organized in cooperation with the RIKILT Institute of food safety, Wageningen, Netherlands. The design of the study in cooperation with the RIKILT Institute of food safety, the study performance in Freising-Weihenstephan and sampling of blood, hair, urine and different tissues at slaughter was part of this PhD thesis.

In planning such animal experiments there are many factors that have to be considered especially concerning the number of animals included in the trial. The guidelines for that trial was to have four different groups. Regarding the present conditions at the Versuchsstation Veitshof, Freising, Germany it was possible to include 20 animals in the trial. The trial was permitted by the Regierung von Oberbayern (Reference Number 55.2-1-5412531.8-102-07). A statistical report required for the application for that animal trial demonstrates that by including two control groups and two groups treated with different doses of hormones, 5 animals per group are adequate for statistical analysis.

So, 20 healthy, 5-7 weeks old Holstein Friesian calves were separated into four groups of five animals each (n=5).

The hormone mix for this pour on study contained 25 mg Oestradiolbenzoate (Intervet, Boxmeer, Netherlands), 60 mg Testosteronedecanoate (Organon, Oss, Netherlands) and 60 mg Testosteronecypionate (Organon). The hormone mix was applied in two different ways: via injection (one animal per group) or via pour on (four animals per group). To ensure the transit of the hormone mix through the skin, following carriers were used: Ivomec (Merial BV, Amstelveen, Netherlands), Dimethylsulfoxid (DMSO) (Sigma-Aldrich, Zwijndrecht, Netherlands), Miglyol 840 (Dynamit Nobel, Germany) und Diethylenglycolmonobutylether (DEGMBE) (Merck, Amsterdam, Netherlands). For injection arachide oil served as carrier.

The first group served as untreated control group (C). In the second group the animals were treated only with the different carriers three times in weekly application (CC). In the third group the animals were treated once with the different carrier hormone mixes (T1). The animals of the fourth group were treated with the different carrier hormone mixes three times in weekly application (T3). To treat the animals via pour on, they were shaven on the back from neck to tail. For treatment 10 mL of hormone carrier mix were administered at the shaved region. The injected animals were treated intramuscular in the neck. Table 1 shows a scheme of the different treatment groups.

Table 1: Treatment scheme of the pour on animal trial

| Group | Animal | Treatment | Group | Animal | Treatment |
|---|---|---|---|---|---|
| C | 1 | none | T1 | 11 | Injection hormones |
| | 2 | none | | 12 | DEGMBE+hormones |
| | 3 | none | | 13 | DMSO+hormones |
| | 4 | none | | 14 | IVOMEC+hormones |
| | 5 | none | | 15 | Miglyol+hormones |
| CC | 6 | Injection Arachide oil | T3 | 16 | Injection hormones |
| | 7 | DEGMBE | | 17 | DEGMBE+hormones |
| | 8 | DMSO | | 18 | DMSO+hormones |
| | 9 | IVOMEC | | 19 | IVOMEC+hormones |
| | 10 | Miglyol | | 20 | Miglyol+hormones |

Blood samples were taken at eight different time points: Predose samples were taken after study start without prior treatment. Further samples were taken at day 2, day 7, day 14, day 21, day 35, day 63, and day 91 of treatment. Blood samples were taken as described previously [51].

During the treatment period the animals were weighted 4 times (before treatment and at days 28, 63, and 91 of treatment).

The animal attendance and blood sampling were done at the Veitshof, Institute of Physiology, Technical University of Munich, Freising-Weihenstephan, Germany. The animals were housed and fed according to practice.

### SARM study on Macaca fascicularis

The aim of this study was to test the possibility of finding gene expression biomarkers for the treatment with a new SARM. Therefore samples for gene expression analyses were taken from an animal trial that was organized within the second phase clinical trial by TAP Pharmaceuticals, Chicago, USA.

In that trial 24 male cynomolgus monkeys (Macaca fascicularis) were separated into four groups of six animals each. All animals were 5-6 years old, skeletally mature and had an average body weight of $6 \pm$ kg. The treatments were control: oral vehicle, Testo: 3.0 mg/kg Testosteronenanthate as Testoviron®-depot-250 (Schering, Berlin, Germany) dosed biweekly by intramuscular injection, SARM1: 1 mg/kg SARM LGD2941 daily, oral and SARM10: 10 mg/kg SARM LGD2941 daily, oral. The oral vehicle control and the SARM were dosed once daily for 90 days.

Whole blood samples were taken at three time points. Predose samples were taken after study start without prior treatment. Further samples were taken at day 16 and day 90 of treatment. Duplicate blood samples were taken as described previously [52].

The animal attendance and blood sampling were done by Covance Laboratories GmbH (Münster, Germany) and was conducted with permission from the local veterinary authorities and in accordance with accepted standards of Human Animal Care.

## 2.2   RNA Extraction and Quality Determination

RNA from blood samples was extracted using the PAXgene Blood RNA Kit (Qiagen, Hilden, Germany). This system is designed for the stabilization and extraction of RNA from human whole blood. Before using it for Macaca fascicularis or cattle it had to be tested, if this system also works for these species. Due to the high homology between human and primate blood the PAXgene system worked well for Macaca fascicularis. The cellular composition of bovine blood differs from that of human blood. There are for example no reticulocytes present, because the maturation of bovine erythrocytes completely takes place in the bone marrow. For testing the PAXgene system with bovine blood, several blood volumes were used. The RNA yield was generally lower compared to the RNA extracted from human blood. It was not practicable to use more than 2.5 ml bovine blood, because by using higher volumes, the stabilizer was not able to lyse all cells and so erythrocytes accumulated and RNA extraction was not possible. Therefore RNA extraction from bovine blood was performed according to the manufacturer´s instructions.

There is no information available about the extraction of RNA from bovine vaginal epithelial cells. The literature only describes how to extract RNA from vaginal epithelial cells obtained from euthanized mice. Our intention was to extract RNA from vaginal epithelial cells of living cattle. A method to get vaginal epithelial cells is to take vaginal smear containing keratinized epithelial cells. The high amount of cervical mucus present in bovine vagina at the end of the estrous cycle was problematic for sampling but by using a sterilized spoon sampling could be successfully performed.

For RNA extraction several systems were tested including phenol based methods and kits using silica membranes. After comparing RNA yield and RNA quality, the method of choice was peqGold Tri-Fast (PeqLab Biotechologies) based on the manufacturer´s instructions.

To quantify the amount of total RNA extracted, optical density (OD) was measured with the Biophotometer (Eppendorf, Hamburg, Germany) or with the NanoDrop 1000 (PeqLab Biotechnologies) for each sample. RNA purity was calculated with the $OD_{260/280}$ ratio.

RNA integrity and quality control was performed via automated capillary electrophoresis in the 2100 Bioanalyzer (Agilent Technologies, Palo Alto, USA). Eukaryotic total RNA Nano Assay (Agilent Technologies) was taken for sample analysis and the RNA Integrity Number (RIN) served as RNA quality parameter. Agilent 2100 Bioanalyzer calculated the RIN value based on a numbering system from 1 to 10 (1 being the most degraded profile, 10 being the most intact) for all samples.

## 2.3 Selection of Target Genes

Candidate genes that might be biomarkers in blood or vaginal smear were chosen by screening the respective literature for steroid related effects on blood or vaginal epithelial cells. Quantified target genes are listed in tables 2-4.

Table 2: Description and accession number of target genes that were quantified in bovine blood [53-57]

| Gene Group | Target Genes | Accession Number |
|---|---|---|
| Reference Genes | Histone H3 | NM_001014389 |
| | Tyrosine 3-monooxygenase/tryptophan 5-monooxygenase activation protein, zeta polypeptide (YWHAZ) | NM_174814 |
| | Glyceraldehyde-3-phosphate dehydrogenase (GAPDH) | U85042 |
| | Ubiquitin 3 (UB3) | Z18245 |
| Steroid receptors | Androgen receptor (AR) | AY862875 |
| | Estrogen receptors $\alpha$ and $\beta$ (ER$\alpha$ and Er$\beta$) | NM_001001443 / NM_174051 |
| | Glucocorticoid receptor $\alpha$ (GR$\alpha$) | AY238475 |
| Apoptosis regulators | TNF receptor superfamily member 6 (Fas) | U34794 |
| | TNF receptor superfamily member 6 ligand (FasL) | XM_584322 |
| | B-cell CLL/lymphoma 2 (BCL-2) | XM_586976 |
| | B-cell lymphoma-extra large (Bcl-Xl) | AF245489 |
| | Tumor necrosis factor $\alpha$ (TNF$\alpha$) | NM173966 |
| | Tumor necrosis factor receptor 1 and 2 (TNFR1, TNFR2) | NM173966 / AF031589 |
| | Caspase 3 and 8 (Casp3, Casp8) | NM_001077840 / DQ319070 |
| Interleukins | Interleukins 1$\alpha$, 1$\beta$, 6, 8, 10, 12B (p40) and 15 (IL-1$\alpha$, IL-1$\beta$, IL-6, IL-8, IL-10, IL-12B, IL-15) | M36182. / M37211 / NM173923 / AF232704 / NM_174088 / NM_174356 / NM_174090 |
| CD Antigens | CD 4, 8 and 14 | NM_001103225 / BC151259 / NM_174008 |

| Growth factors | Insulin-like growth factor 1 (IGF-1) | NM_001077828 |
|---|---|---|
| | Tumor growth factor β (TGFβ) | XM592497 |
| | Interferon gamma (IFN-γ) | NM_174086 |
| others | Inflammatory factor nuclear factor of kappa light polypeptide gene enhancer in B-cells 1 (p105) (NFkB) | NM_001076409 |
| | Actinβ (ACTB) | AY141970 |
| | Actin α 1 (ACTA1) | NP_776650 |
| | Creatine Kinase (CK) | NM_174225 |
| | Adrenergic beta kinase 2 (ADRBK2) | NM_174500 |
| | Major histocompatibility complex class II (MHC II) | NM_001034668 |
| | Jun oncogene (JUN) | NM_001077827 |
| | Estrogen induced transcription factor (EITr) | XR_027981 |
| | Myotrophin (MTPN) | NM_203362 |
| | Tropomodulin 3 (TMOD3) | NM_001075987 |
| | RNA binding protein 5 (RBM5) | NM_001046374 |

Table 3: Description and accession number of target genes that were quantified in bovine vaginal smear [58-65]

| Gene group | Target Genes | Accession Number |
|---|---|---|
| Reference Genes | Histone H3 | NM_001014389 |
| | Tyrosine 3-monooxygenase/tryptophan 5-monooxygenase activation protein, zeta polypeptide (YWHAZ) | NM_174814 |
| Steroid Receptors | Androgen receptor | AY862875 |
| | Estrogen receptor α (ERα) | NM_001001443 |
| | Progesteron receptor (PR) | XM_583951.4 |
| Keratinization Factors | Fibroblast growth factor 7 (FGF7) | XM_869016 |
| | Fibroblast growth factor binding protein (FGFBP) | NM_174337.2 |
| | Cytokeratin 8 (CK8) | BC103339 |
| | Cytokeratin 18 (CK18) | XM_582930 |
| Growth Factors | Epithelial growth factor (EGF) | AY195611.1 |
| | Epithelial growth factor receptor (EGFR) | XM_592211.4 |
| | Insulin like growth factor 1 (IGF-1) | NM_001077828 |
| | Insulin like growth factor 1 receptor (IGF-1R) | X54980 |
| | Insulin like growth factor binding protein (IGFBP3) | NM_174556.1 |
| | Tumor growth factor a (TGFα) | XM_593710.4 |
| | Lactoferrin (LTF) | NM_180998.2 |

| | TNF receptor superfamily member 6 (Fas) | U34794 |
|---|---|---|
| | TNF receptor superfamily member 6 ligand (FasL) | XM_584322 |
| Apoptosis | Caspase 3 and 8 (Casp3, Casp8) | NM_001077840 |
| | | DQ319070 |
| | Tumor necrosis factor α (TNFα) | NM173966 |
| | Tumor necrosis factor receptor 1 (TNFR1) | NM173966 |
| Interleukins | Interleukins 1α, 1β, (IL-1α, IL-1β) | M36182 |
| | | M37211 |
| Oncogens | c jun | AF069515 |
| | c fos | AF069515 |
| Others | Ubiquitin 3 (UB3) | Z18245 |
| | Actinβ (ACTB) | AY141970 |
| | Lingual antimicrobial peptide (LAP) | NM_203435 |

Table 4: Description and accession number of target genes that were quantified in primate blood [54, 55, 66-68]

| Gene Group | Target Genes | Accession Number |
|---|---|---|
| Reference Genes | Actinβ (ACTB) | NM_001101 |
| | Glyceraldehyde-3-phosphate dehydrogenase (GAPDH) | NM_002046 |
| | Ubiquitin 3 (UB3) | NM_021009 |
| Apoptosis regulators | TNF receptor superfamily member 6 (Fas) | NM_000043 |
| | TNF receptor superfamily member 6 ligand (FasL) | NM_000639 |
| | B-cell CLL/lymphoma 2 (BCL-2) | NM_000633 |
| | B-cell lymphoma-extra large (Bcl-Xl) | NM_138578 |
| | Tumor necrosis factor α (TNFα) | NM_000594 |
| | Tumor necrosis factor receptor 1 and 2 (TNFR1, TNFR2) | NM_001065 |
| | | NM_001066 |
| | Caspase 3 and 8 (Casp3, Casp8) | NM_004346 |
| | | NM_001228 |
| | CD 30 ligand (CD30L) | NM_001244 |
| Interleukins | Interleukins 1β, 2, 4, 6, 10, 12B (p40), 13 and 15 (IL-1β, IL-2, IL-4, IL-6, IL-10, IL-12B, IL-13, IL-15) | NM_000576 |
| | | NM_000586 |
| | | NM_172348 |
| | | NM_000600 |
| | | NM_000572 |
| | | NM_002187 |
| | | NM_002188 |
| | | NM_172174 |

| | | |
|---|---|---|
| CD Antigens | CD 4, 8, 11b, 14, 20, 25 and 69 | NM_000616 |
| | | NM_001768 |
| | | NM_000632 |
| | | NM_000591 |
| | | NM_021950 |
| | | NM_000417 |
| | | NM_001781 |
| Growth factors | Insulin-like growth factor 1 receptor (IGF-1R) | NM_000875 |
| | Tumor growth factor β (TGFβ) | NM_000660 |
| inflammatory factors | Inflammatory factor nuclear factor of kappa light polypeptide gene enhancer in B-cells 1 (p105) (NFκB) | NM_003998 |
| | NFκB inhibitor (IκB) | NM_020529 |
| reticulocyte genes | Haemoglobin alpha (α-globin) | NM_000517 |
| | Haemoglobin beta (β-globin) | NM_000518 |
| | Acid phosphatase 1 (αCP1) | NM_006196 |
| | Upstream transcription factor 1 (USF-1) | NM_007122 |
| | Transcription factor CP2 (CP2) | NM_005653 |
| other genes | Androgen receptor (AR) | NM_000044 |
| | Tumor necrosis factor β (TNFβ) | NM_000595 |
| | CD27 ligand (CD27L) | NM_001252 |

## 2.4 Specific Primer Design

All bovine primers were designed using published bovine nucleic acid sequences of GenBank (http://www.ncbi.nlm.nih.gov/entrez/query.fcgi). There are almost no nucleic acid sequences available for *Macaca fascicularis*. For primer design human sequences of the target genes were used. Primer design using human sequences was not possible for all target genes. In this case homologue sequence parts between different species, like mouse, rat and humans were detected using the sequence alignment tool (bl2seq) of the National Center for Biotechnology Information (NCBI) and primer pairs were designed including sequences of the detected homologue parts.

Primer design and optimization was done with primer design program of MWG Biotech (MWG, Ebersberg, Germany) and primer3 (http://frodo.wi.mit.edu/cgi-bin/primer3/primer3_www.cgi) with regard to primer dimer and self-priming formation. Newly designed primers were ordered and synthesized at MWG Biotech (Ebersberg, Germany). Primer testing was performed with three optional samples and a no template control (NTC contains only RNAse free water). To determine the optimal annealing temperature for each primer set a temperature gradient PCR was done.

## 2.5 Two-Step RT-qPCR Analysis

For the studies on bovine tissues, two step RT-qPCR was performed.

### *RNA Reverse Transcription*

Constant amounts of 500 ng or 1 µg RNA were reverse transcribed respectively to cDNA using the following RT master mix: 12 µL 5×Buffer (Promega, Mannheim, Germany), 3 µL Random Hexamer Primers (50 mM; Invitrogen, Carlsbad, USA), 3 µL dNTP Mix (10 mM; Fermentas, St Leon-Rot, Germany) and 200U of MMLV H-Reverse Transcriptase (Promega) according to the manufacturer's instructions.

### qPCR Analysis

To analyze gene expression of candidate genes, RT-qPCR analysis was done with the iQ5 (Bio-Rad, Munich, Germany), using MESA GREEN qPCR MasterMix Plus for SYBR® Assay w/fluorescein Kit (Eurogentec, Cologne, Germany) by a standard protocol, recommended by the manufacture.

With the kit a PCR master mix was prepared as follows: For one sample it is 7.5 µL MESA GREEN 2x PCR Mix, 1.5 µL forward primer (10 pmol/µl), 1.5 µL reverse primer (10 pmol/µl) and 3 µL RNAse free water. For qPCR analysis 1.5 µL cDNA was added to 13.5 µL PCR master mix (total PCR mix: 15µL). qPCR was performed in 96 Well Plates (Eppendorf) and pipetting was done by the epMotion 5075 (Eppendorf).

The following real-time PCR cycling protocol was employed for all investigated factors: denaturation for 5 min at 95°C, 40 cycles of a two segmented amplification and quantification program (denaturation for 3 s at 95°C, annealing for 10 s at primer specific annealing temperature listed in table 1), a melting step by slow heating from 60 to 95°C with a dwell time of 10 s and continous fluorescence measurement. Threshold cycle (Ct) and melting curves were acquired by using the iQ5 Optical System software 2.0 (Bio-Rad). Only genes with clear melting curves were taken for further data analysis. Samples that showed irregular melting peaks were excluded from the quantification procedure.

## 2.6 One-Step RT-qPCR Analysis

For the SARM Study, one step quantitative real time RT-PCR analysis was used, which was performed using SuperScript III Platinum SYBR Green One-Step qPCR Kit (Invitrogen) by a standard protocol, recommended by the manufacture. With the kit a PCR master mix was prepared as follows: For one sample it is 5 µL 2x SYBR Green Reaction Mix, 0.5 µL forward primer (10 pmol/µL), 0.5 µL reverse primer (10 pmol/µL) and 0.2 µL SYBR Green One-Step Enzyme Mix (Invitrogen). 6.2 µL of the PCR master mix was filled in the special 100 µL tubes and 3.8 µL RNA (concentration 1 ng/µL respectively 10 ng/µL) was added (total PCR mix: 10µL). Tubes were closed, placed into the Rotor-Gene 3000 and Analysis Software v6.0 was started (Corbett Life Science, Sydney, Australia). The following uniform one-step RT-qPCR temperature cycling program was used for all genes: Reverse transcription took place at 55°C for 10 min. After 5 min of denaturation at 95°C, 40 cycles of real-time PCR with 3-segment amplification were performed consisting of 15 s at 95°C for denaturation, 30 s at primer dependent temperature for annealing and 20 s at 68°C for polymerase elong ation. The melting step was then performed with slow heating starting at 60°C w ith a rate of 0.5°C per second up to 95°C with continuous measurement of fluoresce nce.

Threshold cycle (Ct) and melting curves were acquired by using the *"Comparative quantitation"* and *"Melting curve"* program of the Rotor-Gene 3000 Analysis software v6.0. Only genes with clear melting curves were taken for further data analysis. Samples that showed irregular melting peaks were excluded from the quantification procedure.

## 2.7 Data Analysis and Statistics

### *Haemogram in bovine studies*
Significant changes of the amount of the different blood cells between the treatment groups were determined using an unpaired t-test. Results with $p < 0.05$ were considered as statistically significant.

### *Weight gain in bovine studies*
For the pour on study, significant changes of carcass weight and the weight gain of the different weighting time points relatively to the beginning of the trial was done by comparing the treatment groups to the control group using an unpaired t-test. Results with $p < 0.05$ were considered as statistically significant.

### *Statistical analysis of gene expression data*
Statistical description of the expression data as well as statistical tests were produced with Sigma Stat for the bovine studies and with SAS v. 9.1.3 for Windows (SAS Institute, Cary, USA) for the SARM study. Since the amplification efficiency was not known, the assumption of identical amplification efficiency 100% was made, allowing more simple quantification model [69].

The Ct values of each gene were averaged by arithmetic mean for each animal. The obtained mean Ct values were then translated to normalized expression quantities using two reference genes (RG) in form of normalization index. The normalization index was calculated as an arithmetic mean of the Ct values of the two RG:

$$\text{normalization index} = \text{mean}(Ct_{RG1}, Ct_{RG2}) \qquad (1)$$

Then, the expression of every target gene was calculated relatively to the expression of the RG as:

$$\text{normalized expression} = 2^{\text{reference index}} / 2^{Ct\ \text{target gene}}, \qquad (2)$$

where the 2 represents the 100% amplification efficiency.

For quantification of gene expression in blood samples the normalized expressions of the treatment timepoints were divided with the normalized expressions of the baseline (predose), generating the expression ratio R as:

$$R_{\text{timepoint/baseline}} = \text{normalized expression}_{\text{timepoint}} / \text{normalized expression}_{\text{baseline}} \qquad (3)$$

The expression ratio R for blood and the normalized expression for vaginal smear was then analysed statistically using the t-test. Results with $p < 0.05$ were considered as statistically significant.

### *Principal components analysis (PCA)*

To disclose multivariate response to the treatment, the method of principal components analysis (PCA) was employed using GenEx v. 4.3.x (MultiD Analyses AB, Gothenburg, Sweden). PCA involves a mathematical procedure that transforms a number of variables (here normalized expression values) into a smaller number of uncorrelated variables called principal components. By this the dimensionality of the data is reduced to a number of dimensions that can be plotted in a scatter plot, here two dimensions. The first principal component accounts for as much of the variability in the data as possible, and each succeeding component accounts for as much of the remaining variability as possible. Normalized expression values of all responding genes were taken as the initial variables and reduced to two principal components only, facilitating thus resolution of treatment clusters in the scatter plot.

### *Hierarchical Cluster Analysis*

Another method for visualizing treatment patterns based on multivariate data is hierarchical cluster analysis. The hierarchical order is represented by a tree dendogram, in which related samples are more closely together than samples that are more different [70, 71]. Hierarchical clustering was employed using GenEx v. 4.3.6 (MultiD Analyses AB).

# 3 Results and Discussion

## 3.1 Anabolics study on Nguni Cattle

### *Haemogram*

The haemograms indicate that the animals were healthy. The white blood cell count and the amount of lymphocytes, monocytes, eosinophil, and basophil granolucytes ranged in physiological levels with no significant differences between treatment group and control (p-values are listed in table 5). Therefore significant changes in mRNA expression in blood can be interpreted as real changes in gene expression and are not due to changes in the blood cell, especially the mRNA expressing white blood cells.

Table 5: List of p-values for the regulation of the amount of the different blood cells

| Timepoint | white blood cell count | lymphocytes | monocytes | eosinophils | basophils |
|---|---|---|---|---|---|
| Predose | 0.5347 | 0.9263 | 0.1273 | 0.1914 | 0.1691 |
| Day 2 | 0.2827 | 0.8051 | 0.8979 | 0.3663 | - |
| Day 16 | 0.9310 | 0.7601 | 0.0848 | 0.3551 | 0.3927 |
| Day 39 | 0.3758 | 0.5106 | 0.4026 | 0.0690 | 0.8353 |

### *RNA integrity*

Good RNA quality is important for the overall success of RNA based analysis methods like real time RT-qPCR [72-75]. The RNA degradation level was determined using the lab-on-a-chip technology of the Agilent Bioanalyzer (Agilent Technologies).

The mean (±SD) RIN value of the blood samples was 8.3 ± 0.3 indicating fully integer total RNA.

The RIN value of the vaginal smear samples was only 4.5 ± 2.02 (mean ± SD). The relatively low RNA quality could be due to the fact that cells found in the vaginal smear are detached, keratinized and partly degraded. Another reason for the low RNA quality results can be RNases present in the vaginal flora. Due to the low RNA quality, the validation of qPCR assays was problematic. Primer pairs had to be designed resulting in PCR products with a length of about 100 bp. This is recommended by Fleige et al for RNA with low quality [76, 77]. Following these guidelines RT-qPCR for 29 genes (27 candidate genes and 2 reference genes) could be successfully established.

### RT-qPCR results and data analysis

*Blood*

Significant regulation of gene expression of the treatment group compared to the control group could be identified for IL-6, MHC II, CK, MTPN and RBM5 after 2 days (figure 3), for GRα, ERα, Fas and IL-1α after 16 days (figure 4) and for ACTB, GRα, IL-1α and IL-1β after 39 days of treatment (figure 5). The resulting p-values and the regulation ratio between control and treatment group are listed in table 6.

Table 6: Significant mRNA expression changes. P-values and x-fold regulation between steroid treatment and control group.

| Gene Group | Gene | Timepoint | p-value | x-fold regulation |
|---|---|---|---|---|
| Steroid receptors | GR | Day 16 | 0.0159 | 1.597 |
|  |  | Day 39 | 0.0273 | 1.345 |
|  | ER | Day 16 | 0.0106 | 1.509 |
| Apoptosis regulators | Fas | Day 16 | 0.0463 | 1.978 |
| Interleukins | IL-1 | Day 16 | 0.0108 | 2.268 |
|  |  | Day 39 | 0.0364 | 1.65 |
|  | IL-1 | Day 39 | 0.0412 | 1.475 |
|  | IL-6 | Day 2 | 0.0125 | 0.434 |
| Others | MHCII | Day 2 | 0.0219 | 0.682 |
|  | CK | Day 2 | 0.0046 | 0.637 |
|  | MTPN | Day 2 | 0.0129 | 0.621 |
|  | RBM5 | Day 2 | 0.0353 | 0.637 |
|  | ACTB | Day 39 | 0.0095 | 1.345 |

In the box-whisker plots, differences in gene expression of the control group compared to baseline can be observed. This reflects the natural variability of the non-induced expression in each studied subject.

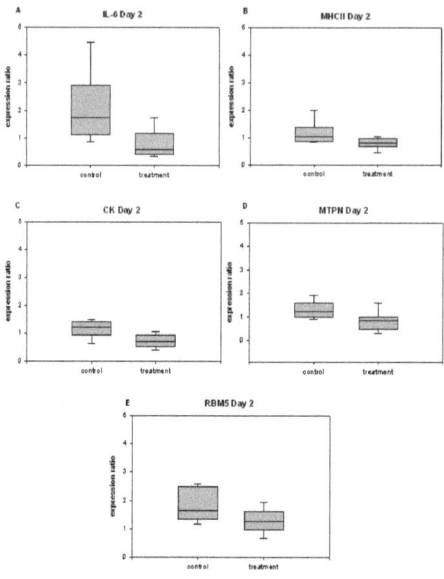

Figure 3: Significant regulations for IL-6 (A), MHCII (B), CK (C), MTPN (D), and RBM5 (E) between control and treated samples after 2 days of treatment.

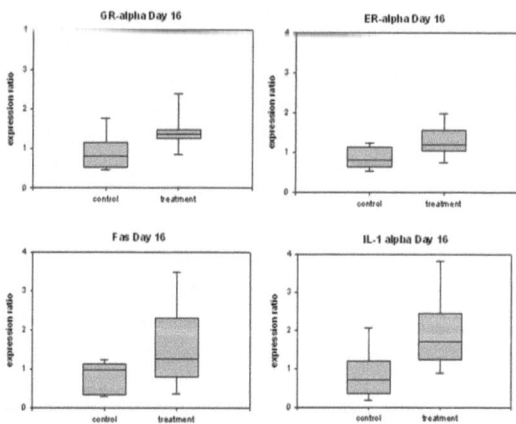

Figure 4: Significant regulations for GRα (A), ERα (B), Fas (C) and IL-1α (D) between control and treated samples after 16 days of treatment.

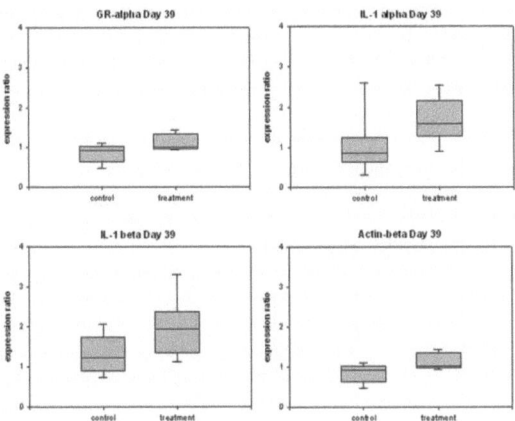

Figure 5: Significant regulations for GRα (A), IL-1α (B), IL-1β (C) and ACTB (D) between control and treated samples after 39 days of treatment.

The number of quantified genes is yet too less to draw conclusions on the different pathways, but anyhow first physiological interpretations can be made and genes that might act as potential biomarkers could be identified.

The steroid receptors GRα and ERα show an up-regulation in the treatment group compared to the control. GRα is up-regulated at day 16 and day 39, whereas ERα is only up-regulated at day 16. Trenbolone acetate has an antiglucocorticoid effect via binding to GR [78-80]. It is already shown that steroid hormones influence the mRNA expression of their receptors in different tissues [65, 81-83]. The applied hormone combination acts via GRα and ERα. The up-regulation of these receptors indicates stimulation of the expression of these receptors in white blood cells.

The interleukins IL-1α and IL-1β are up-regulated. IL-1α is up-regulated at day 16 and day 39, whereas IL-1β is only regulated after 39 days of treatment. IL-1α and IL-1β are produced by macrophages, monocytes and dendritic cells. During infection they induce the release of other cytokines. The expression of IL-1β can be induced by IL-1α. This could be an explanation why IL-1α is up-regulated after 16 days of treatment whereas IL-1β is only up-regulated after 39 days of treatment [84, 85].

PCA is a technique used to reduce multidimensional data sets to lower dimensions for analysis. This statistical method was used to determine whether there is a clustering between control and treatment group. Figure 6 was obtained by plotting all samples of the two groups in the different time points by their two principal components obtained from the 11 regulated genes. Each group was marked by a color. Blue crosses represent samples of the control group and red triangles show the samples of the treatment group. At day 2 and 16 of treatment it can be observed that both groups arrange together and that a difference between control and treatment group can be monitored.

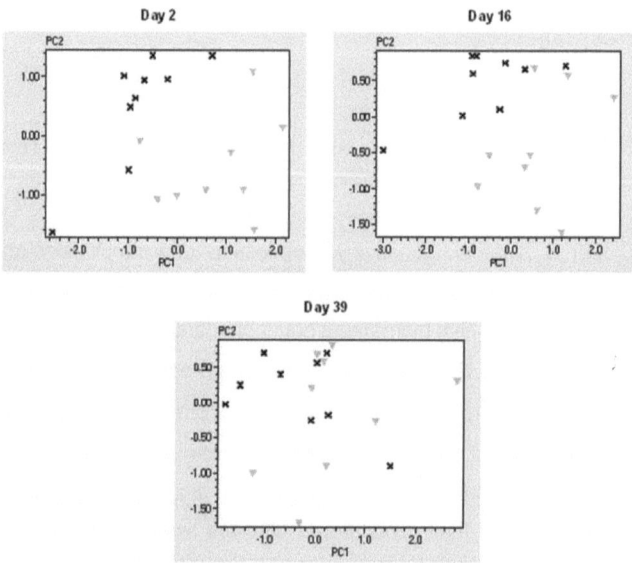

Figure 6: PCA for the eleven regulated genes GR-α, ER-α, Fas, IL-1α, IL-1β, IL-6, MHCII, CK, MTPN, RBM5 and Actin-β at the three different treatment time points. Animals of the control groups are represented by black crosses and animals of the treatment group are represented by grey triangles.

*Vaginal smear*

The steroid receptor ERα showed a significant down-regulation after two days of treatment (p=0.046). Hormones regulate the concentrations of their receptor proteins either by regulating the transcription of the receptor gene or by regulating the stability of the receptor mRNA [86]. The observed down-regulating effect of estrogens on the estrogen receptor was already reported for vaginal cells of mice and rats [87, 88].

The pro-inflammatory interleukins IL-1α (p=0.016) and IL-1β (p=0.005) were both down-regulated after 39 days of treatment.

The keratinization factor CK8 (p=0.003) was significantly down-regulated after two days of treatment. The growth factors FGF7 (p=0.009), EGF (p=0.005), EGFR (p=0.5×10$^{-4}$), TGFα (p=0.5 ×10$^{-3}$), IGF-1R (p=0.007) and LTF (p=0.031) were significantly regulated, whereas EGF, TGFα, IGF-1R were down-regulated at day 2 and FGF7 and EGFR were up-regulated after 16 days and LTF was up-regulated after 39 days of treatment.

At the end of the estrous cycle estrogen levels are high. At this phase vaginal epithelium proliferates, the epithelial cells keratinize and get detached [89].

The keratinization factor CK8 is preferentially expressed in epithelial cells, e.g. in vaginal epithelium. In mice it was already shown that estrogens down-regulate the mRNA expression of this factor [90]. Factors that are involved in the stimulation of the proliferation of epithelial cells are the growth factors FGF7, EGF and EGFR [91-93]. FGF7 and EGF stimulate epithelial growth in vaginal epithelium in mice [94-96]. Both factors were up-regulated after 16 days of treatment. The expression of EGFR was down-regulated after two days of treatment. The effect of estrogens on mRNA expression of these three factors was already shown in mice vaginal epithelial cells [95, 97-99]. The regulation of the growth factors IGF-1R and LTF also goes in line with effects of estrogens that could already be shown in mice. Miyagawa et al. (2004) reported, that the mRNA expression of members of the IGF family is regulated by diethylstilbestrol, a synthetic nonsteroidal estrogen [100]. In this study the down-regulating effect of estrogens on IGF-1R could also be observed. Sato et al. (2004) demonstrated that neonatal exposure of mice with diethylstilbestrol results in an up-regulation of EGF and LTF [99, 101].

It is already known that estrogens stimulate LTF mRNA expression in uterine tissue [64, 102] and that LTF is present at various stages of the estrous cycle in human uterus and vaginal epithelium [64, 103, 104]. This study shows that LTF mRNA expression is increased by estrogen treatment in the bovine vaginal epithelium. The expression of LAP, another defensin was not influenced by the treatment.

Most effects on mRNA expression shown in this study were already obvious in mice and rats. This indicates that the effect of estrogen on the vaginal epithelium is highly conserved. In the 1950s Edgren et al. (1957, 1959) reported that androgens inhibit vaginal effects of estrogens like keratinization of the vaginal epithelium [105, 106]. This study indicates that trenbolone acetate does not show this antagonistic effect.

The oncogene c jun showed a down-regulation at day 2 (p=0.005). Furthermore ACTB (down-regulation at day 2, p=0.007) and UB3 (down-regulation at day 2, p=0.018, and day 16, p=0.001) were significantly regulated. The expression ratios of all regulated genes are listed in table 7.

Table 7: Significant expression changes. Fold regulations between treatment and control group of the significant regulated genes at the three treatment time points.

| Gene Group | Gene | Day 2 | Day 16 | Day 39 |
|---|---|---|---|---|
| Steroid receptors | ERα | 0.59 | | |
| Keratinization factors | CK8 | 0.42 | | |
| Growth factors | FGF7 | | 2.6 | |
| | EGFR | 0.36 | | |
| | EGF | | 2.79 | |
| | IGF-1R | 0.63 | | |
| | TGFα | 0.25 | | |
| | LTF | | | 4.35 |
| Interleukins | IL-1α | | | 0.34 |
| | IL-1β | | | 0.2 |
| Oncogenes | c jun | 0.61 | | |
| Others | ACTB | 0.46 | | |
| | UB3 | 0.64 | 0.31 | |

The second aim of this study was to investigate whether the observed changes of mRNA expression could act as biomarkers to develop a screening method for the combination of trenbolone acetate plus estradiol.

PCA is a technique used to reduce multidimensional data sets to lower dimensions for analysis. This statistical method was used to determine whether there is a clustering between control and treatment group. Figure 7 was obtained by plotting all samples of the two groups in the different time points by their two principal components obtained from the 13 regulated genes. Some genes showed no significant regulation, but showed a trend to be regulated (p<0.1) Therefore PCA was also done by plotting all samples of the two groups in the different time points by their two principal components obtained from all 27 measured candidate genes (Figure 8).

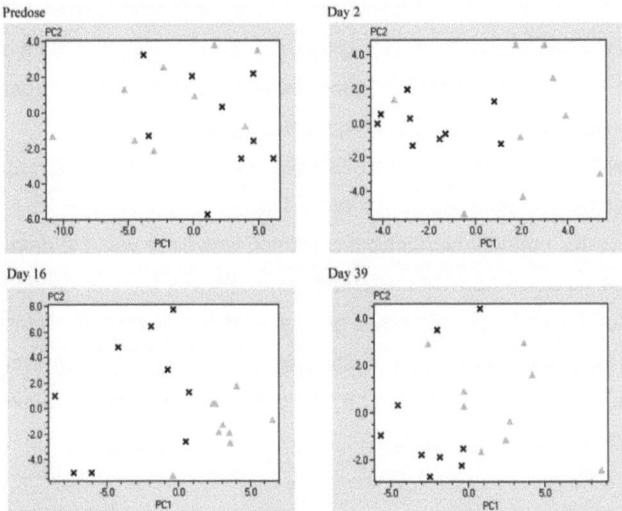

Figure 7: Principal components analysis (PCA) for the thirteen significantly regulated genes at the four different sampling time points. Animals of the control groups are represented by black crosses and animals of the treatment group are represented by grey triangles.

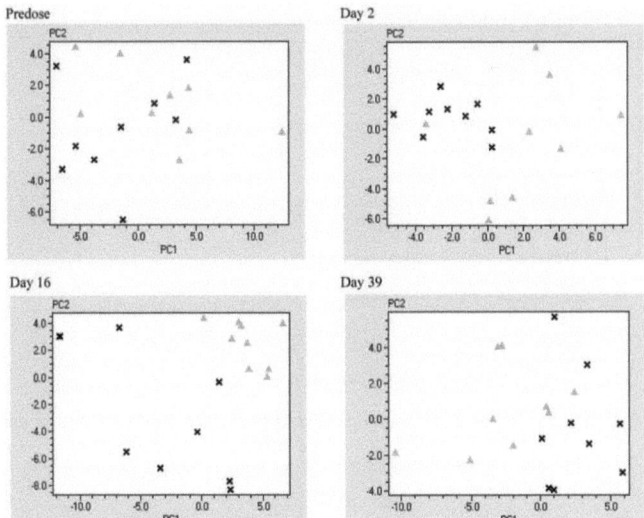

Figure 8: Principal components analysis (PCA) for all 27 measured candidate genes at the four different sampling time points. Animals of the control groups are represented by black crosses and animals of the treatment group are represented by grey triangles.

At all three treatment time points both groups arrange together and a difference between control and treatment group can be monitored. Before treatment the groups show no difference in gene expression of analyzed target genes. This effect is better visible using all 27 quantified genes.

Another biostatistical method to visualize wether the groups arrange together is Hierarchical Cluster Analysis. To verify if the effect observed by PCA is also visibly by using this method, hierarchical clustering was done with the data of the day 16 samples obtained from all measured genes (Figure 9). The dendogram shows a clear separation between the two groups by showing two main branches. The one above only represents control samples. The other one represents treatment samples exept of sample control 6. Performed as treatment screening this sample would be a false positive one.

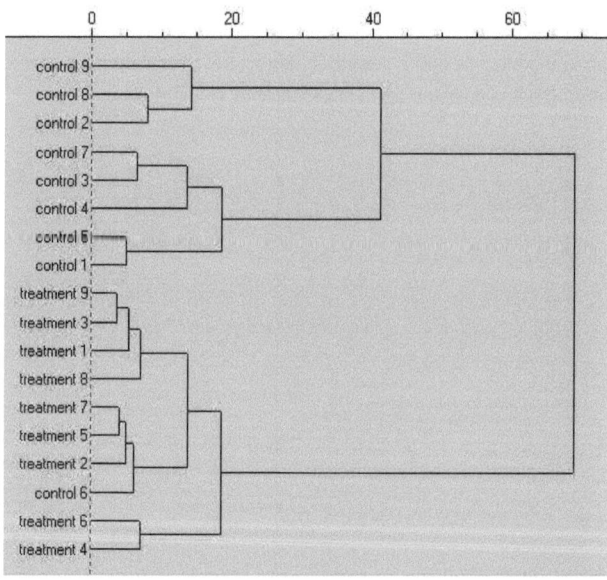

Figure 9: Dendogram for all 27 measured candidate genes at sampling time point day 16.

## 3.2 Pour on anabolics study in veal calves

### *RNA Integrity*
Good RNA quality is important for the overall success of RNA based analysis methods like real time RT-qPCR [74, 75, 107, 108]. The RNA degradation level was determined using the lab-on-a-chip technology of the Agilent Bioanalyzer (Agilent Technologies). The RIN value of the blood samples was 8.5 ± 0.4 (mean ± SD) indicating well intact RNA.

### *Primer testing*
Primer pairs of 32 genes were successfully used in RT-qPCR analysis to get single peaks and uniform melting curves.

### *RT-qPCR results and data analysis*
The carrier control group (CC) showed no significant differences in gene expression compared to the untreated control group and hence the two groups were layed together as one control group of 10 animals for further analyses.

There were no significant differences in the expression of measured target genes 2 and 7 days after treatment start. The steroid receptors ERα, ERβ and GRα were significantly down-regulated in the T3 group, whereas GRα was regulated 14 (p=0.006), ERβ 63 (p=0.031) and ERα 63 (p=0.054) and 91 (p=0.003) days after treatment start. The apoptosis regulators FasL and TNFα were significantly down-regulated in the T3 group, whereas FasL was only regulated 63 days (p=0.050) and TNFα showed a significant regulation 14 (p=0.0005), 21 (p=0.002) and 63 (p=0.004) days after treatment start. The pro-inflammatory factor IL-12B was significantly down-regulated in the T3 group 63 days after treatment start (p=0.010). The transcription factor NFκB showed a significant down-regulation in the T1 (p=0.011) and the T3 group (p=0.012) 63 days after treatment start. CD4 was significantly down-regulated in the T3 group 35 days after treatment start (p=0.025). ACTB was significantly up-regulated in the T3 group 91 days after treatment start (p=0.035) and UB 3 was significantly down-regulated in the T3 group 35 days after treatment start (p=0.045). Table 8 shows the x-fold regulations of all significantly regulated genes at each time point.

Table 8: x-fold regulations of all significantly regulated genes at all timepoints.

| Gene group | Gene | Treatment Group | Day 14 | Day 21 | Day 35 | Day 63 | Day 91 |
|---|---|---|---|---|---|---|---|
| Steroid hormone receptors | ERα | T3 | | | | 0.75 | 0.75 |
| | ERβ | T3 | | | | 0.52 | |
| | GRα | T3 | 0.46 | | | | |
| Apoptosis regulators | FasL | T3 | | | | 0.64 | |
| Pro-inflammatory factors | TNFα | T3 | 0.66 | 0.83 | | 0.76 | |
| | IL-12B | T3 | | | | 0.43 | |
| Transcription factors | NFκB | T1 | | | | 0.81 | |
| | NFκB | T3 | | | | 0.73 | |
| CD Antigen | CD4 | T3 | | | 0.62 | | |
| Others | ACTB | T3 | | | | | 1.23 |
| | UB 3 | T3 | | | 0.76 | | |

The number of quantified genes was yet too less to draw conclusions on the different pathways, but anyhow first physiological declarations can be made and genes that could act as potential biomarkers could be identified.

The mRNA expression of the steroid receptors ERα, ERβ and GRα was significantly down-regulated. It is already shown that steroid hormones influence the mRNA expression of their receptors in different tissues [65, 81, 109, 110], either by regulating the transcription of their receptor gene or by regulating the stability of the receptor mRNA [111].

The significantly down-regulated apoptosis factors TNFα and FasL belong to the TNF Family [112] and induce apoptosis by binding to the death receptors TNFR1, TNFR2 or Fas. The down-regulation of these apoptosis regulators suggest that the immune response is suppressed by the treatment with the used hormone cocktail. This effect seems to be induced by testosterone, because it has already been proven that testosterone has a suppressive effect on the immune system [113, 114].

The pro-inflammatory factor IL-12B was significantly down-regulated in the T3 group 63 days after treatment start. IL-12B – a subunit of IL12 – is mainly produced by monocytes, dendritic cells and activated macrophages. It promotes IFNγ production by CD4 positive T-cells and stimulates proliferation and cytotoxic activity of T-cells and natural killer cells.

### Differences in weight gain

The CC group showed no significant differences in weight gain and carcass weight compared to the untreated control group and hence the two groups were layed together as one control group of 10 animals for further statistical analyses.

No differences in weight gain between the two treatment groups could be observed 28 days after treatment start. 63 days after treatment start the difference in weight gain between the control and the T3 group is not significant but a trend is visible $p<0.1$. At the end of the trial (day 91) the difference in weight gain between the control and the T3 group was significant, whereas there was no mentionable difference between the control and the T1 group. Regarding carcass weight, it can be observed that the T1 group shows a trendly increase compared to the control group ($p = 0.1$) and difference of the T3 group compared to the control group increased significantly ($p = 0.01$).

Differences in weight gain and carcass weight are shown in figure 10 and 11.

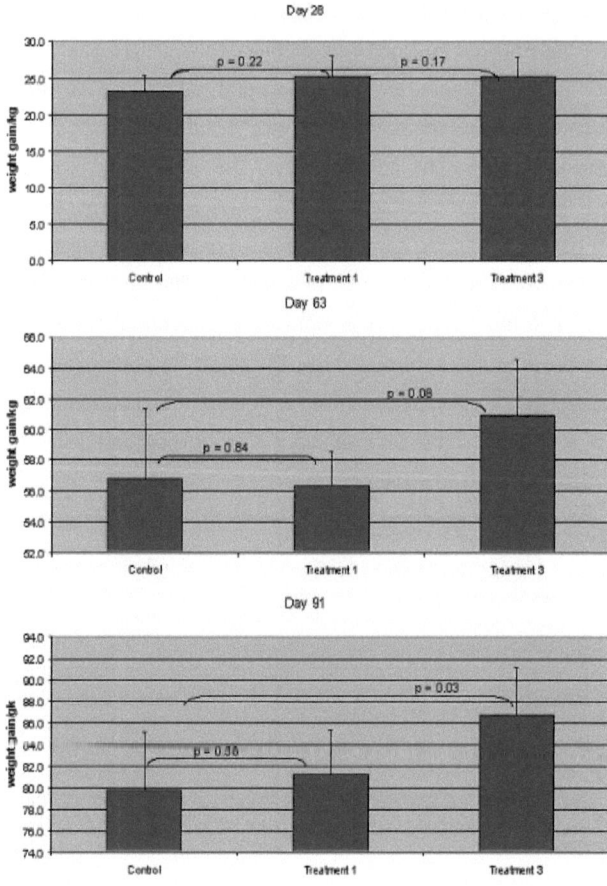

Figure 10 : Differences in live weight gain after 28, 63 and 91 days of treatment

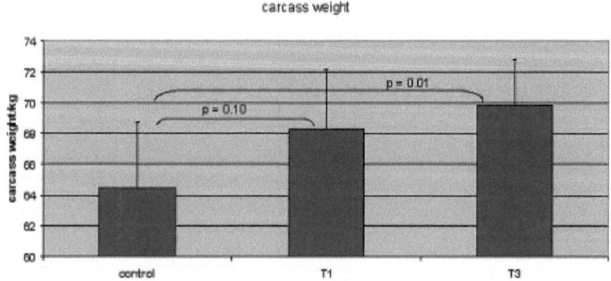

Figure 11 : Differences in carcass weight

The trendly differences in weight gain in the T3 group occur primarily 63 days after treatment start. This is in line with the fact that most differences in gene expression could be observed 63 days after the beginning of treatment. The treatment with anabolic hormones via pour on seems to have no significant effect without being repeated. Another conclusion is that a hormone depot is built which releases the hormones stepwise. This goes in line with observations done by Rattenberger et al. (1993) who could measure hormone residues of diethylstilbestrol and nortestosterone in urine of calves treated with these hormones via pour on even 138 days after treatment [115].

## 3.3  SARM Study on *Macaca fascicularis*

### RNA Integrity
The mean (±SD) RIN value of the blood samples were 7.5 (± 4.8) at predose, 8.5 (± 5.0) on day 16 and 7.7 (± 4.2) at day 90 indicating well intact RNA.

### Primer testing
Primer pairs of 40 genes were successfully used in RT-qPCR analysis to get single peaks and uniform melting curves.

### RT-qPCR results and data analysis
In this study changes of gene expression in blood cells caused by treatment with LGD2941 or testosterone were evaluated in order to describe physiological effects and to find potential biomarkers for the treatment with AR ligands.

Significant down-regulation of gene expression of the treatment groups compared to the control group could be identified for IL-15 ($p=0.0093$) and TNFR2 ($p<0.0001$) after 16 days (Figure 12) and for IL-15 ($p=0.0498$), CD30L ($p=0.0435$), Fas ($p=0.0032$), TNFR1 ($p=0.0308$) and TNFR2 ($p<0.0001$) after 90 days of treatment. Significant up-regulation of gene expression of the treatment groups compared to the control group could be observed for IL-12B ($p=0.0240$) after 90 days of treatment (Figure 13, 14).

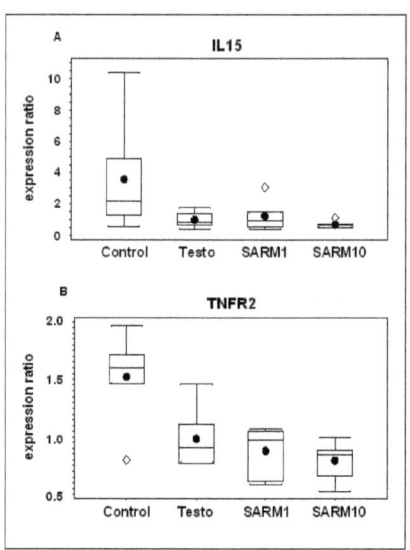

Figure 12: Significant regulation for IL-15 (A) and TNFR2 (B) between control and treated samples after 16 days of treatment.

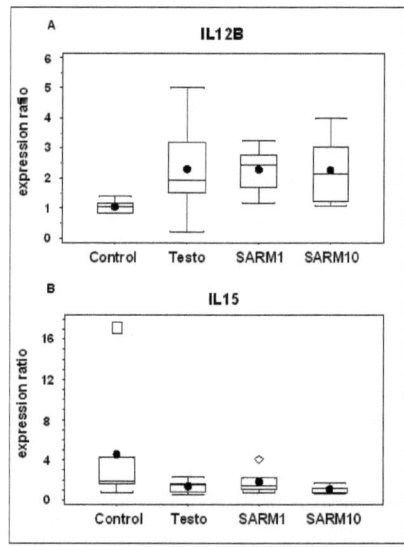

Figure 13: Significant regulation for the proinflammatory interleukins IL-12B (A) and IL-15 (B), between control and treated samples after 90 days of treatment.

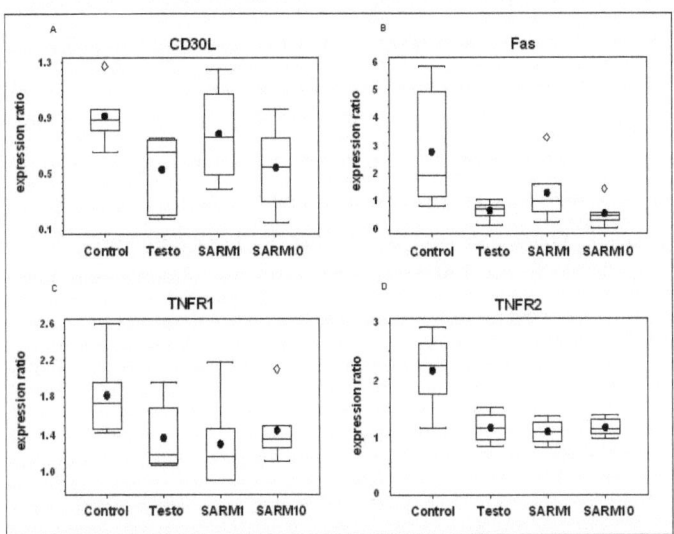

Figure 14: Significant regulation for the apoptosis regulators CD30L (A), Fas (B), TNFR1 (C) and TNFR2 (D) between control and treated samples after 90 days of treatment.

Regarding the Box-whisker plots it can be monitored that the statistical variance in the control group is higher than in the treatment groups. The reason for this could be the natural variability of the non induced expression in each studied subject. Suppression of gene expression by an external stimulus like treatment with testosterone or the SARM reduces natural variability of gene expression.

The main physiological effect that could be observed in this study is the down-regulation of various apoptotic marker genes in all three treatment groups. This is shown by the significant regulation ($p<0.05$) of the apoptosis receptors Fas, TNFR1, TNFR2 and the apoptosis ligand CD30L. All regulated apoptosis factors belong either to the TNF Family (CD30L) or to the TNF-Receptor Family (TNFR1, TNFR2, Fas) [116]. The down-regulation of these apoptosis regulators suggest that the immune response is suppressed by the treatment with testosterone and the SARM. This is consistent with the fact that testosterone has a suppressive effect on the immune system [117, 118]. If the physiological effects of testosterone and the SARM are compared it became obvious that the SARM is similarly active to natural androgens.

PCA is a technique used to reduce multidimensional data sets to lower dimensions for analysis. Figure 15 was obtained by plotting all samples of the four treatment groups by their two principal components obtained from the six responder genes. Blue dots represent samples of the control group, light green dots show the testosterone group, olive dots represent the SARM1 group and the red dots display the SARM10 group. A distinct control group can be seen, showing that there was a multitranscriptional response to the treatment by any of the three drugs. In addition, the SARM1 neighbors to the control group, creating thus a transition to the Testosterone group and the SARM10 group.

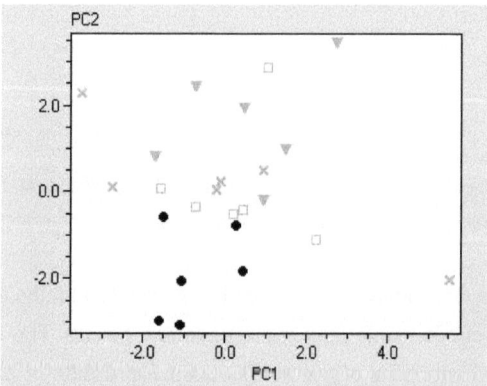

Figure 15: PCA for the six regulated genes IL-12B, IL-15, CD30L, Fas, TNFR1 and TNFR2 in the control group (black dots) the testosterone treated group (grey cross) the low dosed SARM group (grey squares) and the high dosed SARM group (grey triangle).

To verify if there is any correlation of the different regulated genes, the six responder genes were clustered by PCA (figure 16). Red dots show apoptosis regulators and black spots display the interleukins. The TNF receptors cluster very closed together, so that the two spots representing these factors are difficult to separate. This indicates that the two TNF receptors might be coregulated. The other genes do not show any coherence.

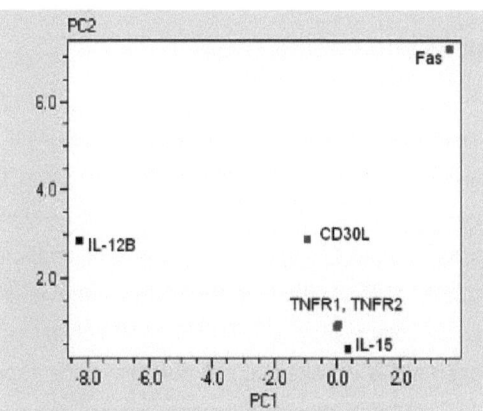

Figure 16: PCA for the regulated genes in all four groups. Black spots show the interleukins and grey spots show the apoptosis regulators

The second aim of this study was to find potential biomarkers for the use of the SARM. Regarding PCA it can be postulated that the regulated genes found in this study can act as first biomarker candidates for the development of a screening pattern in whole blood.

# 4  Conclusions and Perspectives

In all three animal trials included in this thesis the potential of gene expression analysis for developing a new screening method to trace the use of anabolic steroid hormones is examined. Combined with biostatistical methods, like PCA or hierarchical cluster analysis this approach seems to be auspicious.

Although the quality of RNA obtained from bovine vaginal smear is poor, gene expression data in combination with PCA or hierarchical cluster analysis show promising results for the development of potential gene expression biomarkers. Both biostatistical methods show a clear clustering of the treatment groups. The disadvantage of this matrix is, that vaginal smear is only available from female animals. Regarding this, blood samples display a better matrix, because blood can be taken from the living animal independent of gender.

In all three animal trials, changes in gene expression could be quantified in blood samples. Comparing the results obtained from the bovine animal trials it could be observed that only a few genes show expression changes in both studies. Genes like the steroid receptors ERα and GRα are regulated in both experiments but are regulated in different directions. The only gene whose mRNA expression is similarly regulated in both studies is ACTD. Comparing the results obtained from blood samples of the bovine trials and the SARM study it could be shown that only Fas and IL-12B are regulated in both systems, whereas the genes are regulated in different directions. Figure 17 presents an overview of all regulated genes obtained from the three animal trials.

These results indicate that the influence of anabolic steroid hormones on gene expression in blood is species specific and dependent on breed, age, application method and possibly the applied hormone.

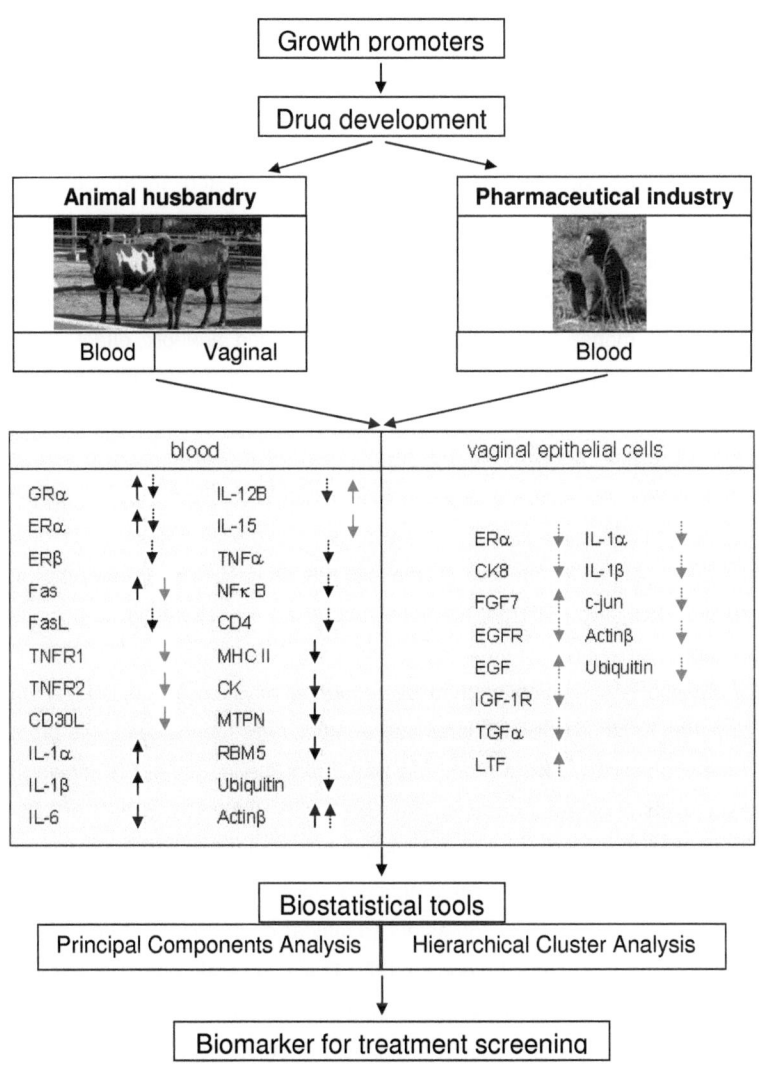

Figure 17: Schematic description of the regulated genes obtained in this thesis.
The direction of regulation is described by arrows. ↓ describes down-regulation and ↑ describes up-regulation. The different tissues and studies are marked by color and line. Black arrows describe results obtained from blood samples in the study on Nguni cattle, Black dotted arrows describe results obtained from blood samples of the pour on study, grey arrows describe results obtained from blood samples of the SARM study and grey dotted arrows describe results obtained from vaginal smear samples of the study on Nguni cattle.

In all three studies the method of PCA was employed to prove if a treatment pattern is visible. Using results obtained from blood samples in the trial on Nguni heifers and the SARM study, clustering of the treatment groups is visible.

Regarding the pour on trial, most changes in gene expression are present 63 days after treatment start, but only in the three times treated group. Obtained gene expression changes are minimal so that biostatistical tools like PCA or hierarchical cluster analysis do not show any successful clustering of the animals.

Although weight gain results of the pour on trial indicate that application of anabolic hormones via pour on show the intended anabolic effect, the use of anabolic steroid hormones by this application method seems not to be detectable on the level of gene expression in blood. This indicates first problems of using gene expression analysis for the development of a new method to screen for a broad range of anabolic agents independent of their way of application. Nevertheless, to determine if the use of gene expression changes to develop a biomarker pattern for the use of anabolic agents is still promising, other tissues, like liver, kidney or different hormone responsive muscles have to be taken into account.

To develop a screening method by regarding physiological effects of anabolic hormones in blood, the additional use of other *omic* technologies like proteomics or metabolomics will be a promising way. From the literature it is known that the blood levels of different proteins, like IGF-1, IGF-1BP3, GH, HDL-C or LDL-C are changed by the treatment of anabolic agents. Regarding metabolomics, most is known about the influence of β-agonists on blood metabolites. Levels of Nτ-methylhistidine, creatinine, non-esterified fatty acids or natural catecholamines are influenced by the use of β-agonists [37]. There have also been efforts to detect perturbations in the metabolic profile after the administration of steroid hormones to reveal the illict application as growth promoters. Blood metabolites like creatinine or creatine kinase, which are associated with muscle function, or the plasma urea levels, which are known to be an early indicator for anabolic effects in cattle, could serve as potential biomarkers for treatment screening [37].

The combination of transcriptomics, proteomics and metabolomics with special biostatistical methods, like hierarchical cluster analysis, canonical correlation analysis and linear or multiple discriminant analysis will be a new perspective for developing a new screening method to trace the abuse of anabolic steroid hormones [37].

# 5 References

1. Meyer, H. H. D. Biochemistry and physiology of anabolic hormones used for improvement of meat production. APMIS. 2001; 109: 1-8

2. Lange, I. G., Daxenberger, A., and Meyer, H. H. Hormone contents in peripheral tissues after correct and off-label use of growth promoting hormones in cattle: effect of the implant preparations Filaplix-H, Raglo, Synovex-H and Synovex Plus. APMIS. 2001; 109: 53-65

3. Meyer, H. H. D. Biochemistry and physiology of anabolic hormones used for improvement of meat production. APMIS. 2001; 109: 1-8

4. Hau, M. Regulation of male traits by testosterone: implications for the evolution of vertebrate life histories. Bioessays. 2007; 29: 133-144

5. Saudan, C., Baume, N., Robinson, N., Avois, L., Mangin, P., and Saugy, M. Testosterone and doping control. Br.J.Sports Med. 2006; 40 Suppl 1: i21-i24

6. Lange, I. G., Daxenberger, A., and Meyer, H. H. Hormone contents in peripheral tissues after correct and off-label use of growth promoting hormones in cattle: effect of the implant preparations Filaplix-H, Raglo, Synovex-H and Synovex Plus. APMIS. 2001; 109: 53-65

7. Moran, C., Quirke, J. F., Prendiville, D. J., Bourke, S., and Roche, J. F. The effect of estradiol, trenbolone acetate, or zeranol on growth rate, mammary development, carcass traits, and plasma estradiol concentrations of beef heifers. J.Anim Sci. 1991; 69: 4249-4258

8. Lange, I. G., Daxenberger, A., and Meyer, H. H. Hormone contents in peripheral tissues after correct and off-label use of growth promoting hormones in cattle: effect of the implant preparations Filaplix-H, Raglo, Synovex-H and Synovex Plus. APMIS. 2001; 109: 53-65

9. Daxenberger, A., Ibarreta, D., and Meyer, H. H. D. Possible health impact of animal oestrogens in food. Human Reproduction Update. 2001; 7: 340-355

10. Swan, S. H., Liu, F., Overstreet, J. W., Brazil, C., and Skakkebaek, N. E. Semen quality of fertile US males in relation to their mothers' beef consumption during pregnancy. Human Reproduction. 2007; 22: 1497-1502

11. Maume, D., Deceuninck, Y., Pouponneau, K., Paris, A., Le Bizec, B., and Andre, F. Assessment of estradiol and its metabolites in meat. APMIS. 2001; 109: 32-38

12. Meyer, H. H. D. Biochemistry and physiology of anabolic hormones used for improvement of meat production. APMIS. 2001; 109: 1-8

13. Lange, I. G., Daxenberger, A., and Meyer, H. H. Hormone contents in peripheral tissues after correct and off-label use of growth promoting hormones in cattle: effect of the implant preparations Filaplix-H, Raglo, Synovex-H and Synovex Plus. APMIS. 2001; 109: 53-65

14. Moran, C., Quirke, J. F., Prendiville, D. J., Bourke, S., and Roche, J. F. The effect of estradiol, trenbolone acetate, or zeranol on growth rate, mammary development, carcass traits, and plasma estradiol concentrations of beef heifers. J.Anim Sci. 1991; 69: 4249-4258

15. Schilt, R., Groot, M. J., Berende, P. L., Ramazza, V., Ossenkoppele, J. S., Haasnoot, W., Van Bennekom, E. O., Brouwer, L., and Hooijerink, H. Pour on application of growth promoters in veal calves: analytical and histological results. Analyst. 1998; 123: 2665-2670

16. Stephany, R. W. Hormones in meat: different approaches in the EU and in the USA. APMIS Suppl. 2001; S357-S363

17. Walsh, M. C., Hunter, G. R., and Livingstone, M. B. Sarcopenia in premenopausal and postmenopausal women with osteopenia, osteoporosis and normal bone mineral density. Osteoporos.Int. 2006; 17: 61-67

18. Morley, J. E. Anorexia, sarcopenia, and aging. Nutrition. 2001; 17: 660-663

19. Morley, J. E., Kim, M. J., and Haren, M. T. Frailty and hormones. Rev.Endocr.Metab Disord. 2005; 6: 101-108

20. Morley, J. E., Baumgartner, R. N., Roubenoff, R., Mayer, J., and Nair, K. S. Sarcopenia. J.Lab Clin.Med. 2001; 137: 231-243

21. Marcell, T. J. Sarcopenia: causes, consequences, and preventions. J.Gerontol.A Biol.Sci.Med.Sci. 2003; 58: M911-M916

22. Morley, J. E., Kim, M. J., and Haren, M. T. Frailty and hormones. Rev.Endocr.Metab Disord. 2005; 6: 101-108

23. Lopez, F. J. New approaches to the treatment of osteoporosis. Curr.Opin.Chem.Biol. 2000; 4: 383-393

24. Deschenes, M. R. Effects of aging on muscle fibre type and size. Sports Med. 2004; 34: 809-824

25. Doherty, T. J. Invited review: Aging and sarcopenia. J.Appl.Physiol. 2003; 95: 1717-1727

26. Negro-Vilar, A. Selective androgen receptor modulators (SARMs): a novel approach to androgen therapy for the new millennium. J.Clin.Endocrinol.Metab. 1999; 84: 3459-3462

27. Negro-Vilar, A. Selective androgen receptor modulators (SARMs): a novel approach to androgen therapy for the new millennium. J.Clin.Endocrinol.Metab. 1999; 84: 3459-3462

28. Meyer, H. H. D. and Hoffmann, S. Development of a sensitive microtitration plate enzyme-immunoassay for the anabolic steroid trenbolone. Food Addit.Contam. 1987; 4: 149-160

29. Meyer, H. H. D., Rinke, L., and Dursch, I. Residue screening for the beta-agonists clenbuterol, salbutamol and cimaterol in urine using enzyme immunoassay and high-performance liquid chromatography. J.Chromatogr. 4-5-1991; 564: 551-556

30. Scippo, M. L., Degand, G., Duyckaerts, A., Maghuin-Rogister, G., and Delahaut, P. Control of the illegal administration of natural steroid hormones in the plasma of bulls and heifers. Analyst. 1994; 119: 2639-2644

31. van Ginkel, L. A. Immunoaffinity chromatography, its applicability and limitations in multi-residue analysis of anabolizing and doping agents. Journal of Chromatography. 4-5-1991; 564: 363-384

32. Becker, K. F., Metzger, V., Hipp, S., and Hofler, H. Clinical proteomics: new trends for protein microarrays. Curr.Med.Chem. 2006; 13: 1831-1837

33. Ludwig, J. A. and Weinstein, J. N. Biomarkers in cancer staging, prognosis and treatment selection. Nat.Rev.Cancer. 2005; 5: 845-856

34. Ludwig, J. A. and Weinstein, J. N. Biomarkers in cancer staging, prognosis and treatment selection. Nat.Rev.Cancer. 2005; 5: 845-856

35. Ilyin, S. E., Belkowski, S. M., and Plata-Salaman, C. R. Biomarker discovery and validation: technologies and integrative approaches. Trends Biotechnol. 2004; 22: 411-416

36. Zhang, X., Li, L., Wei, D., Yap, Y., and Chen, F. Moving cancer diagnostics from bench to bedside. Trends Biotechnol. 2007; 25: 166-173

37. Riedmaier, I., Becker, C., Pfaffl, M. W., and Meyer, H. H. D. The use of *omic* technologies for biomarker development to trace functions of anabolic agents. J.Chrom.A. 2009;

38. Beato, M., Chavez, S., and Truss, M. Transcriptional regulation by steroid hormones. Steroids. 1996; 61: 240-251

39. Edwards, D. P. Regulation of signal transduction pathways by estrogen and progesterone. Annu.Rev.Physiol. 2005; 67: 335-376

40. Picard, D., Kumar, V., Chambon, P., and Yamamoto, K. R. Signal transduction by steroid hormones: nuclear localization is differentially regulated in estrogen and glucocorticoid receptors. Cell Regul. 1990; 1: 291-299

41. Rehberger, P., Rexin, M., and Gehring, U. Heterotetrameric structure of the human progesterone receptor. Proc.Natl.Acad.Sci.U.S.A. 9-1-1992; 89: 8001-8005

42. Rexin, M., Busch, W., Segnitz, B., and Gehring, U. Structure of the glucocorticoid receptor in intact cells in the absence of hormone. J.Biol.Chem. 5-15-1992; 267: 9619-9621

43. Segnitz, B. and Gehring, U. Subunit structure of the nonactivated human estrogen receptor. Proc.Natl.Acad.Sci.U.S.A. 3-14-1995; 92: 2179-2183

44. Beato, M. Gene regulation by steroid hormones. Cell. 2-10-1989; 56: 335-344

45. Beato, M., Chavez, S., and Truss, M. Transcriptional regulation by steroid hormones. Steroids. 1996; 61: 240-251

46. Griekspoor, A., Zwart, W., Neefjes, J., and Michalides, R. Visualizing the action of steroid hormone receptors in living cells. Nucl.Recept.Signal. 2007; 5: e003

47. Weigel, N. L. and Moore, N. L. Kinases and protein phosphorylation as regulators of steroid hormone action. Nucl.Recept.Signal. 2007; 5: e005

48. Ing, N. H. Steroid hormones regulate gene expression posttranscriptionally by altering the stabilities of messenger RNAs. Biol.Reprod. 2005; 72: 1290-1296

49. Meisonnier, E. and Mitchell-Vigneron, J.; Anabolics in Animal Production - Public health aspects, analytical methods and regulation; 1983; 92-9044-118-6

50. Riedmaier, I., Tichopad, A., Reiter, M., Pfaffl, M. W., and Meyer, H. H. Identification of potential gene expression biomarkers for the surveillance of anabolic agents in bovine blood cells. Anal.Chim.Acta. 4-6-2009; 638: 106-113

51. Riedmaier, I., Tichopad, A., Reiter, M., Pfaffl, M. W., and Meyer, H. H. Identification of potential gene expression biomarkers for the surveillance of anabolic agents in bovine blood cells. Anal.Chim.Acta. 4-6-2009; 638: 106-113

52. Riedmaier, I., Tichopad, A., Reiter, M., Pfaffl, M. W., and Meyer, H. H. D. Influence of testosterone and a novel SARM on gene expression in whole blood of *macaca fascicularis*. J.Steroid Biochem.Mol.Biol. 2009;

53. Ing, N. H. Steroid hormones regulate gene expression posttranscriptionally by altering the stabilities of messenger RNAs. Biol.Reprod. 2005; 72: 1290-1296

54. Cutolo, M., Capellino, S., Montagna, P., Ghiorzo, P., Sulli, A., and Villaggio, B. Sex hormone modulation of cell growth and apoptosis of the human monocytic/macrophage cell line. Arthritis Research & Therapy. 2005; 7: R1124-R1132

55. Huber, S. A., Kupperman, J., and Newell, M. K. Estradiol prevents and testosterone promotes Fas-dependent apoptosis in CD4+ Th2 cells by altering Bcl 2 expression. Lupus. 1999; 8: 384-387

56. Jacobson, J. D. and Ansari, M. A. Immunomodulatory actions of gonadal steroids may be mediated by gonadotropin-releasing hormone. Endocrinology. 2004; 145: 330-336

57. Lehmann, D., Siebold, K., Emmons, L. R., and Muller, H. Androgens inhibit proliferation of human peripheral blood lymphocytes in vitro. Clin.Immunol.Immunopathol. 1988; 46: 122-128

58. Kang, J. S., Lee, B. J., Ahn, B., Kim, D. J., Nam, S. Y., Yun, Y. W., Nam, K. T., Choi, M., Kim, H. S., Jang, D. D., Lee, Y. S., and Yang, K. H. Expression of estrogen receptor alpha and beta in the uterus and vagina of immature rats treated with 17-ethinyl estradiol. J.Vet.Med.Sci. 2003; 65: 1293-1297

59. Ing, N. H. Steroid hormones regulate gene expression posttranscriptionally by altering the stabilities of messenger RNAs. Biol.Reprod. 2005; 72: 1290-1296

60. Pessina, M. A., Hoyt, R. F., Jr., Goldstein, I., and Traish, A. M. Differential regulation of the expression of estrogen, progesterone, and androgen

receptors by sex steroid hormones in the vagina: immunohistochemical studies. J.Sex Med. 2006; 3: 804-814

61. Sato, T., Fukazawa, Y., Ohta, Y., and Iguchi, T. Involvement of growth factors in induction of persistent proliferation of vaginal epithelium of mice exposed neonatally to diethylstilbestrol. Reprod.Toxicol. 2004; 19: 43-51

62. Traish, A. M., Kim, S. W., Stankovic, M., Goldstein, I., and Kim, N. N. Testosterone increases blood flow and expression of androgen and estrogen receptors in the rat vagina. J.Sex Med. 2007; 4: 609-619

63. Miyagawa, S., Suzuki, A., Katsu, Y., Kobayashi, M., Goto, M., Handa, H., Watanabe, H., and Iguchi, T. Persistent gene expression in mouse vagina exposed neonatally to diethylstilbestrol. J.Mol.Endocrinol. 2004; 32: 663-677

64. Teng, C. T. Lactoferrin gene expression and regulation: an overview. Biochem.Cell Biol. 2002; 80: 7-16

65. Pfaffl, M. W., Daxenberger, A., Hageleit, M., and Meyer, H. H. D. Effects of synthetic progestagens on the mRNA expression of androgen receptor, progesterone receptor, oestrogen receptor alpha and beta, insulin-like growth factor-1 (IGF-1) and IGF-1 receptor in heifer tissues. JOURNAL OF VETERINARY MEDICINE SERIES A-PHYSIOLOGY PATHOLOGY CLINICAL MEDICINE. 2002; 49: 57-64

66. McMurray, R. W., Suwannaroj, S., Ndebele, K., and Jenkins, J. K. Differential effects of sex steroids on T and B cells: modulation of cell cycle phase distribution, apoptosis and bcl-2 protein levels. Pathobiology. 2001; 69: 44-58

67. Jacobson, J. D. and Ansari, M. A. Immunomodulatory actions of gonadal steroids may be mediated by gonadotropin-releasing hormone. Endocrinology. 2004; 145: 330-336

68. Lehmann, D., Siebold, K., Emmons, L. R., and Muller, H. Androgens inhibit proliferation of human peripheral blood lymphocytes in vitro. Clin.Immunol.Immunopathol. 1988; 46: 122-128

69. Livak, K. J. and Schmittgen, T. D. Analysis of relative gene expression data using real-time quantitative PCR and the 2(-Delta Delta C(T)) Method. Methods. 2001; 25: 402-408

70. Pfaffl, M. W., Vandesompele, J., and Kubista, M. Real-Time PCR: Current Technology and Applications. Caister Academic Press. London, GB. 2009, 5.

71. Beyene, J., Tritchler, D., Bull, S. B., Cartier, K. C., Jonasdottir, G., Kraja, A. T., Li, N., Nock, N. L., Parkhomenko, E., Rao, J. S., Stein, C. M., Sutradhar, R., Waaijenborg, S., Wang, K. S., Wang, Y., and Wolkow, P. Multivariate analysis of complex gene expression and clinical phenotypes with genetic marker data. Genet.Epidemiol. 2007; 31 Suppl 1: S103-S109

72. Fleige, S., Walf, V., Huch, S., Prgomet, C., Sehm, J., and Pfaffl, M. W. Comparison of relative mRNA quantification models and the impact of RNA integrity in quantitative real-time RT-PCR. Biotechnol.Lett. 2006; 28: 1601-1613

73. Fleige, S. and Pfaffl, M. W. RNA integrity and the effect on the real-time qRT-PCR performance. Mol.Aspects Med. 2006; 27: 126-139

74. Pfaffl, M., Fleige, S., and Riedmaier, I. Validation of lab-on-chip capillary electrophoresis systems for total RNA quality and quality control. Biotechnol & Biotechnol eq. 2008; 22: 829-834

75. Schroeder, A., Mueller, O., Stocker, S., Salowsky, R., Leiber, M., Gassmann, M., Lightfoot, S., Menzel, W., Granzow, M., and Ragg, T. The RIN: an RNA integrity number for assigning integrity values to RNA measurements. BMC Molecular Biology. 2006; 7: 3

76. Fleige, S., Walf, V., Huch, S., Prgomet, C., Sehm, J., and Pfaffl, M. W. Comparison of relative mRNA quantification models and the impact of RNA integrity in quantitative real-time RT-PCR. Biotechnol.Lett. 2006; 28: 1601-1613

77. Fleige, S. and Pfaffl, M. W. RNA integrity and the effect on the real-time qRT-PCR performance. Mol.Aspects Med. 2006; 27: 126-139

78. Bauer, E. R., Daxenberger, A., Petri, T., Sauerwein, H., and Meyer, H. H. Characterisation of the affinity of different anabolics and synthetic hormones to the human androgen receptor, human sex hormone binding globulin and to the bovine progestin receptor. APMIS. 2000; 108: 838-846

79. Meyer, H. H. D. Biochemistry and physiology of anabolic hormones used for improvement of meat production. APMIS. 2001; 109: 1-8

80. Pottier, J., Cousty, C., Heitzman, R. J., and Reynolds, I. P. Differences in the biotransformation of a 17 beta-hydroxylated steroid, trenbolone acetate, in rat and cow. Xenobiotica. 1981; 11: 489-500

81. Pfaffl, M. W., Lange, I. G., and Meyer, H. H. D. The gastrointestinal tract as target of steroid hormone action: quantification of steroid receptor mRNA expression (AR, ERalpha, ERbeta and PR) in 10 bovine gastrointestinal tract compartments by kinetic RT-PCR. The Journal of Steroid Biochemistry and Molecular Biology. 2003; 84: 159-166

82. Reiter, M., Walf, V. M., Christians, A., Pfaffl, M. W., and Meyer, H. H. Modification of mRNA expression after treatment with anabolic agents and the usefulness for gene expression-biomarkers. Anal.Chim.Acta. 3-14-2007; 586: 73-81

83. Toffolatti, L., Rosa, Gastaldo L., Patarnello, T., Romualdi, C., Merlanti, R., Montesissa, C., Poppi, L., Castagnaro, M., and Bargelloni, L. Expression analysis of androgen-responsive genes in the prostate of veal calves treated with anabolic hormones. Domest.Anim Endocrinol. 2006; 30: 38-55

84. Dinarello, C. A. The biological properties of interleukin-1. Eur.Cytokine Netw. 1994; 5: 517-531

85. Dinarello, C. A. The interleukin-1 family: 10 years of discovery. FASEB J. 1994; 8: 1314-1325

86. Ing, N. H. Steroid hormones regulate gene expression posttranscriptionally by altering the stabilities of messenger RNAs. Biol.Reprod. 2005; 72: 1290-1296

87. Kang, J. S., Lee, B. J., Ahn, B., Kim, D. J., Nam, S. Y., Yun, Y. W., Nam, K. T., Choi, M., Kim, H. S., Jang, D. D., Lee, Y. S., and Yang, K. H. Expression of estrogen receptor alpha and beta in the uterus and vagina of immature rats treated with 17-ethinyl estradiol. J.Vet.Med.Sci. 2003; 65: 1293-1297

88. Sato, T., Fukazawa, Y., Ohta, Y., and Iguchi, T. Involvement of growth factors in induction of persistent proliferation of vaginal epithelium of mice exposed neonatally to diethylstilbestrol. Reprod.Toxicol. 2004; 19: 43-51

89. Kronenberg, M. S. and Clark, J. H. Changes in keratin expression during the estrogen-mediated differentiation of rat vaginal epithelium. Endocrinology. 1985; 117: 1480-1489

90. Miyagawa, S., Suzuki, A., Katsu, Y., Kobayashi, M., Goto, M., Handa, H., Watanabe, H., and Iguchi, T. Persistent gene expression in mouse vagina exposed neonatally to diethylstilbestrol. J.Mol.Endocrinol. 2004; 32: 663-677

91. Sato, T., Fukazawa, Y., Ohta, Y., and Iguchi, T. Involvement of growth factors in induction of persistent proliferation of vaginal epithelium of mice exposed neonatally to diethylstilbestrol. Reprod.Toxicol. 2004; 19: 43-51

92. Iguchi, T., Uchima, F. D., and Bern, H. A. Growth of mouse vaginal epithelial cells in culture: effect of sera and supplemented serum-free media. In Vitro Cell Development Biology. 1987; 23: 535-540

93. Iguchi, T., Uchima, F. D., Ostrander, P. L., and Bern, H. A. Growth of normal mouse vaginal epithelial cells in and on collagen gels. Proceedings of the National Academy of Sciences. 1983; 80: 3743-3747

94. Miyagawa, S., Suzuki, A., Katsu, Y., Kobayashi, M., Goto, M., Handa, H., Watanabe, H., and Iguchi, T. Persistent gene expression in mouse vagina exposed neonatally to diethylstilbestrol. J.Mol.Endocrinol. 2004; 32: 663-677

95. Hom, Y. K., Young, P., Thomson, A. A., and Cunha, G. R. Keratinocyte growth factor injected into female mouse neonates stimulates uterine and vaginal epithelial growth. Endocrinology. 1998; 139: 3772-3779

96. Hom, Y. K., Young, P., Wiesen, J. F., Miettinen, P. J., Derynck, R., Werb, Z., and Cunha, G. R. Uterine and vaginal organ growth requires epidermal growth factor receptor signaling from stroma. Endocrinology. 1998; 139: 913-921

97. Sato, T., Fukazawa, Y., Ohta, Y., and Iguchi, T. Involvement of growth factors in induction of persistent proliferation of vaginal epithelium of mice exposed neonatally to diethylstilbestrol. Reprod.Toxicol. 2004; 19: 43-51

98. Hom, Y. K., Young, P., Wiesen, J. F., Miettinen, P. J., Derynck, R., Werb, Z., and Cunha, G. R. Uterine and vaginal organ growth requires epidermal growth factor receptor signaling from stroma. Endocrinology. 1998; 139: 913-921

99. Nelson, K. G., Sakai, Y., Eitzman, B., Steed, T., and McLachlan, J. Exposure to diethylstilbestrol during a critical developmental period of the mouse reproductive tract leads to persistent induction of two estrogen-regulated genes. Cell Growth Differentiation. 1994; 5: 595-606

100. Miyagawa, S., Suzuki, A., Katsu, Y., Kobayashi, M., Goto, M., Handa, H., Watanabe, H., and Iguchi, T. Persistent gene expression in mouse vagina exposed neonatally to diethylstilbestrol. J.Mol.Endocrinol. 2004; 32: 663-677

101. Sato, T., Fukazawa, Y., Ohta, Y., and Iguchi, T. Involvement of growth factors in induction of persistent proliferation of vaginal epithelium of mice exposed neonatally to diethylstilbestrol. Reprod.Toxicol. 2004; 19: 43-51

102. Pentecost, B. T. and Teng, C. T. Lactotransferrin is the major estrogen inducible protein of mouse uterine secretions. J.Biol.Chem. 7-25-1987; 262: 10134-10139

103. Cohen, M. S., Britigan, B. E., French, M., and Bean, K. Preliminary observations on lactoferrin secretion in human vaginal mucus: variation during the menstrual cycle, evidence of hormonal regulation, and implications for infection with Neisseria gonorrhoeae. American Jorunal of Obstetrics and Gynecology. 1987; 157: 1122-1125

104. Kelver, M. E., Kaul, A., Nowicki, B., Findley, W. E., Hutchens, T. W., and Nagamani, M. Estrogen regulation of lactoferrin expression in human endometrium. Am.J.Reprod.Immunol. 1996; 36: 243-247

105. EDGREN, R. A. Oestrogen antagonisms: the effects of testosterone propionate and 17-ethyl-19-nortestosterone on oestrone-induced changes in vaginal cytology. Acta Endocrinol.(Copenh). 1957; 25: 365-370

106. EDGREN, R. A. The modification of estrogen-induced changes in rat vaginas with steroids and related agents. Annals of the New York Academy of Sciences. 11-18-1959; 83: 160-184

107. Fleige, S., Walf, V., Huch, S., Prgomet, C., Sehm, J., and Pfaffl, M. W. Comparison of relative mRNA quantification models and the impact of RNA integrity in quantitative real-time RT-PCR. Biotechnol.Lett. 2006; 28: 1601-1613

108. Fleige, S. and Pfaffl, M. W. RNA integrity and the effect on the real-time qRT-PCR performance. Mol.Aspects Med. 2006; 27: 126-139

109. Reiter, M., Walf, V. M., Christians, A., Pfaffl, M. W., and Meyer, H. H. Modification of mRNA expression after treatment with anabolic agents and the usefulness for gene expression-biomarkers. Anal.Chim.Acta. 3-14-2007; 586: 73-81

110. Toffolatti, L., Rosa, Gastaldo L., Patarnello, T., Romualdi, C., Merlanti, R., Montesissa, C., Poppi, L., Castagnaro, M., and Bargelloni, L. Expression analysis of androgen-responsive genes in the prostate of veal calves treated with anabolic hormones. Domest.Anim Endocrinol. 2006; 30: 38-55

111. Ing, N. H. Steroid hormones regulate gene expression posttranscriptionally by altering the stabilities of messenger RNAs. Biol.Reprod. 2005; 72: 1290-1296

112. Nagata, S. Apoptosis by death factor. Cell. 2-7-1997; 88: 355-365

113. Morell, V. Zeroing in on how hormones affect the immune system. Science. 8-11-1995; 269: 773-775

114. Verthelyi, D. Sex hormones as immunomodulators in health and disease. Int.Immunopharmacol. 2001; 1: 983-993

115. Rattenberger, E., Wnuk, I., and Matzke, P. Pharmacokinetics of des and 19-Nortestosterone in Calves After Spot-on Treatment. Archiv für Lebensmittelhygiene. 1993; 44: 73-76

116. Nagata, S. Apoptosis by death factor. Cell. 2-7-1997; 88: 355-365

117. Morell, V. Zeroing in on how hormones affect the immune system. Science. 8-11-1995; 269: 773-775

118. Verthelyi, D. Sex hormones as immunomodulators in health and disease. Int.Immunopharmacol. 2001; 1: 983-993

# Acknowledgements

First of all I would like to thank Prof. Dr. Dr. Heinrich H.D. Meyer for offering me the opportunity to work in the interesting research field of anabolic steroid hormones at the chair of Physiology and for supervising me during this time. I want to thank him for all the very inspiring discussions concerning my work and for his patience to answer all my questions throughout the time span of my PhD thesis.

Many thanks to PD Dr. Michael W. Pfaffl for supervising me concerning all methodological questions and for the really good working athmosphere.

I want to thank Dr. Ales Tichopad who showed me how to deal with great amounts of data and who taught me statistical thinking.

Special thanks to Azel Swemmer and Dr. Maria Groot without whom the bovine studies included in this thesis could not take place.

I would like to thank Christiane Becker, Christine Fochtmann, Gabriele Jobst and all other participants of the "pour on anabolics" trial who helped me to successfully carry out this trial.

Many thanks to my colleagues and friends Dr. Martina Reiter, Dr. Simone Fleige, Dr. Bettina Griesbeck-Zilch, Dr. Heike Kliem, all my neighbors in the "Denkerzentrale" and to all employees at the Institute for the nice working atmosphere and good collaboration. Special thanks to Christiane Becker and Andrea Hammerle-Fickinger for their friendship and for cheering me up when not everything went the right way.

I want to thank all my friends at home, who tolerated the lack of time present during the last three years.

Many thanks to my parents and siblings for enabling me this education and for supporting me in all its forms

Special thanks goes to my boyfriend Klaus for simply everything.

I am deeply grateful for the financial support of the projects by "TAP Pharmaceuticals", the "Onderstepoort Veterinary Institute, Pretoria" and the "RIKILT Institute of Food Safety".

# Scientific Communication

**Original Publications**

M.W. Pfaffl, S. Fleige, I. Riedmaier
Validation of lab-on-chip capillary electrophoresis systems for total RNA quality and quantity control
*Biotechnology and Biotechnological Equipment*, 2008, 22/3.

I. Riedmaier, C. Becker, M.W. Pfaffl, H.H.D. Meyer
The use of *omic* technologies for biomarker development to trace anabolic hormone functions
*Journal of Chromatography A* (2009), doi:10.1016/j.chroma.2009.01.094

I. Riedmaier, A. Tichopad, M. Reiter, M.W. Pfaffl, H.H.D. Meyer
Influence of testosterone and a novel SARM on gene expression in whole blood of *Macaca fascicularis*
*Journal of Steroid Biochemistry and Molecular Biology*, 2009, 114: 167-173

I. Riedmaier, A. Tichopad, M. Reiter, M.W. Pfaffl, H.H.D. Meyer
Identification of potential gene expression biomarkers for the surveillance of anabolic agents in bovine blood cells
*Analytica Chimica Acta*, 2009; 638: 106-113

I. Riedmaier, M. Reiter, A. Tichopad, M.W. Pfaffl, H.H.D. Meyer
The potential of bovine vaginal smear for biomarker development to trace the misuse of anabolic agents
Submitted: "The Analyst"

**Oral Presentations:**

I. Riedmaier, C. Becker, M.W. Pfaffl, H.H.D. Meyer
The use of transcriptomics for biomarker development to trace anabolic hormone functions
4th International qPCR Symposium, Technische Universität München, 09.03.-13.03.2009, Freising-Weihenstephan, Germany.

**Poster Presentations:**

I. Riedmaier, A. Tichopad, M. Reiter, M.W. Pfaffl, H.H.D. Meyer
Influence of testosterone and a novel SARM on gene expression in whole blood of *Macaca fascicularis*
18th Symposium of the Journal of Steroid Biochemistry and Molecular Biology, 18.09.-21.09. 2008, Seefeld in Tirol, Österreich

I. Riedmaier, M. Reiter, A. Tichopad, M.W. Pfaffl, H.H.D. Meyer
Identification of potential gene expression biomarkers in bovine vaginal smear after application of the anabolic combination trenbolone acetate plus estradiol
18th Symposium of the Journal of Steroid Biochemistry and Molecular Biology, 18.09.-21.09. 2008, Seefeld in Tirol, Österreich

I. Riedmaier, M. Bergmaier, M.W. Pfaffl
Comparison of two available platforms for the determination of RNA quality
4th International qPCR Symposium, Technische Universität München, 09.03.-13.03.2009, Freising-Weihenstephan, Germany.

# Appendix

**Appendix I:**
I. Riedmaier, C. Becker, M.W. Pfaffl, H.H.D. Meyer
The use of *omic* technologies for biomarker development to trace anabolic hormone functions
*Journal of Chromatography A,* 2009, 1216(46): 8192-8199

**Appendix II:**
I. Riedmaier, A. Tichopad, M. Reiter, M.W. Pfaffl, H.H.D. Meyer
Influence of testosterone and a novel SARM on gene expression in whole blood of *Macaca fascicularis*
*Journal of Steroid Biochemistry and Molecular Biology,* 2009, 114: 167-173

**Appendix III:**
I. Riedmaier, A. Tichopad, M. Reiter, M.W. Pfaffl, H.H.D. Meyer
Identification of potential gene expression biomarkers for the surveillance of anabolic agents in bovine blood cells
*Analytica Chimica Acta,* 2009; 638: 106-113

# Appendix I

## Journal of Chromatography A

journal homepage: www.elsevier.com/locate/chroma

Review

# The use of *omic* technologies for biomarker development to trace functions of anabolic agents

Irmgard Riedmaier*, Christiane Becker, Michael W. Pfaffl, Heinrich H.D. Meyer

*Physiology Weihenstephan, Technische Universitaet Muenchen, Weihenstephaner Berg 3, 85354 Freising, Germany*

### ARTICLE INFO

*Article history:*
Available online 5 February 2009

*Keywords:*
Omic technologies
Transcriptomics
Proteomics
Metabolomics
Hormone analysis

### ABSTRACT

The combat against misuse of growth promoting agents is a major topic in agricultural meat production and human sports. In routine screening, hormone residues of all known growth promoting agents are detected by immuno assays or chromatographical methods in combination with mass spectrometry. To overcome the detection by these routine screening methods new xenobiotic growth promoters and new ways of application were developed, e.g. the combination of different agents in hormone cocktails are employed. To enable an efficient tracing of misused anabolic substances it is necessary to develop new screening technologies for a broad range of illegal drugs including newly designed xenobiotic anabolic agents. The use of *omic* technologies like, transcriptomics, proteomics or metabolomics is a promising approach to discover the misuse of anabolic hormones by indirectly detecting their physiological action. With the help of biostatistical tools it is possible to extract the quested information from the data sets retrieved from the *omic* technologies. This review describes the potential of these *omic* technologies for the development of such new screening methods and presents recent literature in this field.

© 2009 Elsevier B.V. All rights reserved.

### Contents

1. Introduction ... 8192
2. Molecular mechanisms of steroid hormone signaling ... 8193
3. Molecular mechanisms of β-agonist signaling ... 8194
4. The use of *omic* technologies for biomarker research ... 8194
   4.1. Transcriptomics ... 8194
   4.2. Proteomics ... 8195
   4.3. Metabolomics ... 8196
5. Bioinformatics ... 8197
6. Conclusions ... 8197
   References ... 8198

### 1. Introduction

Natural steroid hormones are synthesized from cholesterol and they are strongly involved in endocrine and paracrine regulation of growth and differentiation in most tissues. Some steroid hormones, like estradiol or testosterone show anabolic functions by enhancing body protein accretion and mobilizing fat stores, which results in an increased growth rate [1]. These properties are deep-rooted in the evolution of vertebrates. The sex steroids testosterone and estradiol have effects on behavioral, morphological and physiological traits. Estrogens stimulate protein- and mineral retention during pregnancy which is important for the development of the embryo. Testosterone promotes sexual behaviors like courtship, and improves growth of skeletal muscle which is important for defending the territory [1,2].

Steroid hormones participate in the establishment of muscle tissue and bone density. After menopause women and also older men often suffer from a loss in muscle mass (sarcopenia) and bone mineral density (osteoporosis) which may lead to frailty [3–6]. Both conditions are related to the decrease in the endogenous production of anabolic sex hormones, mainly estradiol and testosterone [5]. Men and women suffering from frailty are treated with

* Corresponding author. Fax: +49 8161 714204.
  *E-mail address:* irmgard.riedmaier@wzw.tum.de (I. Riedmaier).

0021-9673/$ – see front matter © 2009 Elsevier B.V. All rights reserved.
doi:10.1016/j.chroma.2009.01.094

testosterone or estradiol but both therapies are associated with negative side effects like skin virilization in women, prostate hypertrophy in men and an increased risk of cancer [7-9]. An alternative to the treatment with natural testosterone or estradiol are synthetic molecules called SARM (selective androgen receptor modulators) and SERM (selective estrogen receptor modulators), which bind to the steroid hormone receptors exhibiting predominantly tissue selective effects [10].

In human sports and agricultural meat producing animals the myotropic, growth promoting properties of steroid hormones are very beneficial. Used orally, the natural steroid hormones testosterone and estradiol are almost inactive. Besides these natural steroids the xenobiotic hormones trenbolone acetate (TBA), zeranol and melengestrol acetate (MGA) were developed by US companies to be used as anabolics in food producing animals. Whereas only MGA is orally active, the other drugs have to be applied by implantation [11].

Besides steroid hormones, the substance group of β-agonists has also been used as growth promoter in animal husbandry and human sports. β-agonists are well known in medicine due to their vasodilative attributes to treat asthma and other pulmonary diseases [12]. A widely spread drug for this application is salbutamol, which is given by the inhaled route to act directly on the smooth muscle cells in the bronchia. Synthetic β-agonists like salbutamol or the orally active clenbuterol are derivates of the adrenal medullary hormone epinephrin and the neurotransmitter norepinephrin, which are the natural agonists of the β-adrenergic receptor [13].

Several studies document the anabolic action of β-agonists in farm animals and also in laboratory animals. The daily weight gain of bulls treated with β-agonists was shown to be significantly higher than that in the untreated control group [14]. Increased growth rates and improved feed conversion could be observed in finishing bulls fed climaterol [15]. Rats fed clenbuterol improved live weight gain and feed efficiency as well as increased muscle mass due to hypertrophy of muscle fibres [16]. Another effect of β-agonists is the degradation of fat stores and hence the increase of the fat to lean meat ratio [17]. Because of that impact besides anabolic steroid hormones, these substances are used as growth promoters in animal husbandry [1,18,19]. Zilpaterol and ractopamine were developed by international companies to modify nutrient partition in food producing animals.

In meat production growth promoters are used to increase productivity and to reduce costs by improving weight gain and feed efficiency [20,21]. The use of growth promoters is approved in some countries like the USA, Canada, Mexico, Australia and South Africa. It has been proven that hormone residues in meat are increased and have adverse side effects to the consumer [20,22-24]. Therefore the use of anabolic agents in meat producing animals and also the import of meat derived from cattle given these substances is forbidden in the EU since 1988 (88/146/EEC). To enforce the EU-directive, permanent surveillance is essential [1,20-22,25,26].

In human sports, the application of anabolic substances to increase muscle performance, called doping, increased in the past 40 years [27-31]. Anabolic agents are not only used by competitive athletes, professional body builders or weight lifters, but more and more by amateurs to improve appearance and body shaping [32]. The World Anti-Doping Agency (WADA) yearly publishes a list of drugs and substance classes that are forbidden to be taken during training and competition [11,32,33]. The large number of doping cases during the Tour de France 2007 showed the importance of improving the screening techniques that can be used in future doping control practice and the requirement to develop new approaches to become more efficient in view of the upcoming new classes of growth promoters.

To uncover the abuse of anabolic agents in animal husbandry and human sports hormone residues are detected using immuno

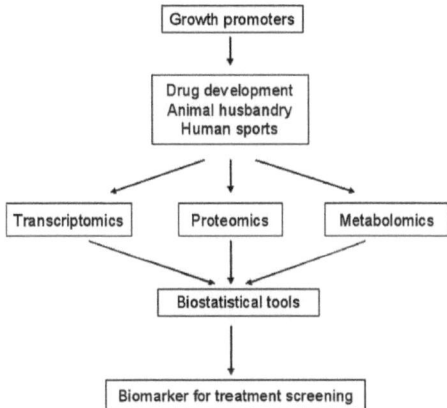

Fig. 1. Scheme of the use of *omic* technologies to trace anabolic hormone functions.

assays or chromatographical methods in combination with mass spectrometry [34-37]. With these methods only known substances can be discovered. To enable an efficient tracing of misused anabolic substances it is necessary to develop new technologies to screen for a broad range of illegal drugs including newly designed xenobiotic anabolic agents.

In molecular medicine, e.g. in cancer research, the development of molecular biomarkers is already a common approach in diagnostics. Plasma biomarkers are developed for prognostic use and tumor biomarkers are used to develop treatment strategies for each individual patient [38,39]. To develop such biomarkers *omic* technologies, like transcriptomics, proteomics and metabolomics are applied [39-41].

The use of such *omic* technologies will be a promising way to develop a biomarker pattern based on physiological changes that are caused after illegal application of anabolic agents (Fig. 1).

This review reflects efforts made during the last two decades in the field of screening for anabolic agents in animal husbandry and describes physiological and molecular effects of anabolic agents on different tissues in order to illustrate the potential of *omic* technologies for the development of reliable molecular biomarkers for anabolic agents. Literature research was done by using common databases for biomedical literature.

## 2. Molecular mechanisms of steroid hormone signaling

Steroid hormone receptors belong to the family of nuclear receptors and show a high affinity to their corresponding hormone [42,43]. They are either localized in the cytoplasm moving to the cell nucleus upon activation or directly in the nucleus waiting for the steroid hormones or active analoga to enter the nucleus and activate them [44]. Steroid receptors consist of different domains like a DNA binding domain, a nuclear localization domain, a ligand binding domain and several transactivation domains [42]. Without a bound ligand the steroid receptors exist as a steroid receptor complex, associated with different heat shock proteins (hsp90, hsp 56, hsp70) and p23 [45-47]. Binding of the ligand results in a conformational change which leads to the dissociation of the HSP-complex from the receptor. After dimerization the receptor binds to specific sequences in the promoter region of steroid hormone regulated genes, called hormone responsive elements (HRE) [42,43,48]. After DNA binding, different coregulators that are needed for

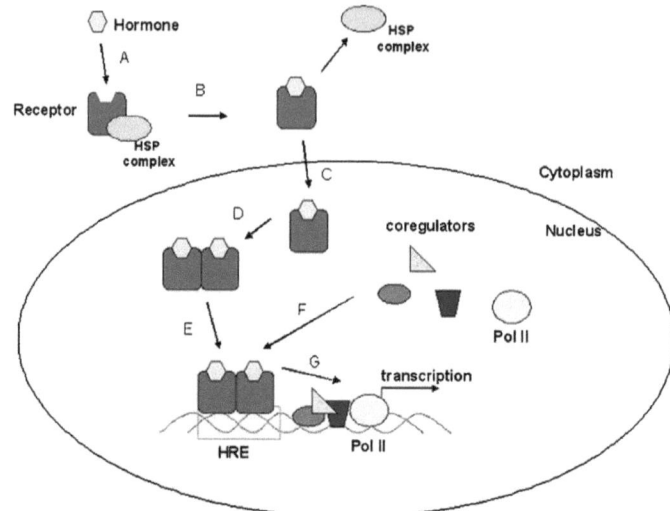

**Fig. 2.** Schematic diagram of the activation of a cytoplasmic steroid hormone receptor. After hormone binding (A) the HSP complex dissociates from the receptor (B), the hormone receptor complex translocates to the nucleus (C), dimerizes (D) and binds to a hormone responsive element (HRE) in the promoter region of a specific gene (E). After binding to the HRE different coregulators of transcription are recruited (F), which are responsible for transcriptional activation [49,50].

transcriptional activation are recruited. These coregulators have different functions. They either enhance or repress transcription through enzymatic activities like acetylation, deacetylation, kinase activity or methylation [49]. These coregulators are for example responsible for chromatin remodeling or the recruitment of RNA polymerase II (Pol II) [50] (Fig. 2). Another possibility of regulating gene transcription by steroid hormones is to influence or recruit other transcription factors like AP1 [51,52] or NFκB [53].

Steroid hormones not solely regulate gene transcription activity but also influence the stability of generated mRNA. They are able to stabilize or destabilize specific mRNA. Most is known about the influence of steroid hormones on the stability of their receptor mRNA. Whereas steroid receptor protein is normally down-regulated by their ligands, the regulation of the stability of steroid receptor mRNA may be positive or negative. Regulation of mRNA stability is not restricted to steroid hormone receptors, other genes are also regulated by similar mechanism [54].

Sex steroid hormone receptor signaling is primarily important in tissues of the reproductive tract like uterus, ovary, vagina, testes or prostate. But also other tissues like muscles, liver, kidney, lung, spleen, blood cells and parts of the gastrointestinal tract express steroid hormone receptors and are influenced by steroid hormones [55–57].

## 3. Molecular mechanisms of β-agonist signaling

As β-adrenergic receptors are present on almost every mammalian cell their agonists exert diverse biochemical effects. β-adrenergic receptors belong to the group of seven-span trans membrane receptors. Physiological mechanisms of β-agonists are mediated by binding of the agonist to the β-adrenergic receptor and the following induction of a G-protein coupled signaling cascade (Fig. 3).

The α-subunit of the G-protein thereby activates the enzyme adenylate cyclase (AC), which produces cyclic adenosine monophosphate (cAMP) as intracellular signaling molecule. After binding to the regulatory subunit of protein kinase C (PKC) cAMP removes the catalytic subunit to enable the enzyme to phosphorylate several intracellular proteins. This phosphorylation can either activate (e.g. hormone sensitive lipase) or deactivate (e.g. acetyl CoA carboxylase) enzymes. PKC also phosphorylates cAMP responsive element binding protein (CREBP), which binds to cAMP responsive elements (CRE) in regulatory regions of genes to stimulate transcription [58,59].

## 4. The use of *omic* technologies for biomarker research

### 4.1. Transcriptomics

The transcriptome is the complete set of RNA transcripts present in a particular cell, and the most prominent candidates investigated in research are the messenger RNA (mRNA), micro-RNA (miRNA), transfer RNA (tRNA), and ribosomal RNA (rRNA). Transcriptomics describes the global study of gene expression at a certain time point for example as a reaction after a specific treatment.

Methods used nowadays for studying transcriptomics are cDNA hybridisation microarrays, conventional RT-PCR and quantitative real-time RT-PCR (qRT-PCR). Microarrays have the advantage that a whole set of genes can be analyzed on one array, but they are not sensitive enough to measure minimal changes in gene expression. Using quantitative RT-PCR genes can only be quantified separately but this method is more sensitive, its dynamic range of quantitation is much wider, it is better reproducible and less expensive than microarray experiments. Another advantage of qRT-PCR is that more biological samples can be measured in one experiment [60,61].

The combination of both, finding biomarker candidate genes using microarrays or exploring the literature and verification of these changes in gene expression using qRT-PCR is a promising way to find gene expression biomarkers.

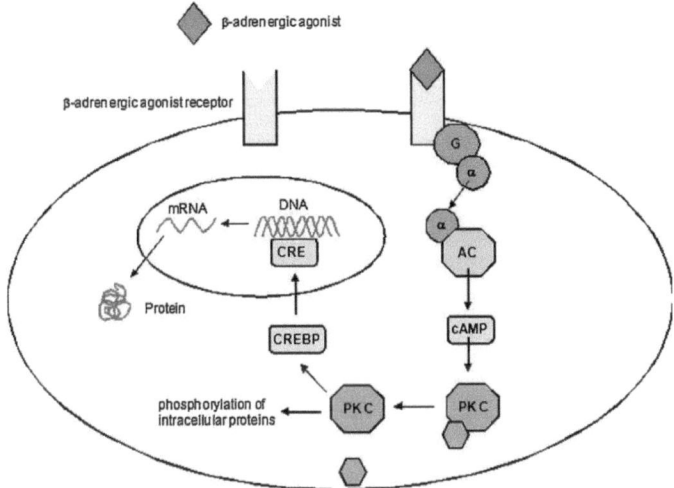

Fig. 3. Signaling cascade leading to the physiological mechanisms of β-adrenergic agonists.

The potential power of gene expression biomarkers for diagnostic use has already been demonstrated in cancer research [41,62–66]. Physiological changes can be quantified on the level of gene expression. Anabolic hormones have several physiological effects and therefore finding gene expression biomarkers could be a promising approach to develop a screening method for the use or misuse of anabolic hormones.

There are numerous reports that steroid hormones and also β-agonists affect gene expression in different organs. Reiter and coworkers [67] quantified changes in mRNA expression for a number of genes in bovine liver, muscle and uterus that are controlled by different xenobiotic anabolic agents and found several regulated genes that could be first candidates for developing gene expression biomarkers.

The influence of steroid hormones on the mRNA expression of several genes could be shown by some research groups in different tissues. Promising candidate genes for the development of a screening method in cattle are IGF-1 in liver and muscle [56,67–69], steroid hormone receptors in various tissues like liver, muscle, uterus, the gastrointestinal tract, kidney, prostate and blood cells [56,57,67,70,71], and various inflammatory, apoptotic and proliferative genes in blood cells [71–73]. β-agonists are known to affect mRNA expression of different muscle proteins like α-actin, myosin or calpastatin in cattle. The mRNA expression of β-adrenergic receptors are also known to be influenced by their ligands [74,75].

Most of these tissues can only be taken after slaughter and so they present no promising tissues for developing a doping screening method in humans. In humans only non-invasive sampling of blood, urine or hair could serve as matrix to find gene expression changes, because they can easily be taken from the individual.

In vivo studies in humans regarding gene expression changes caused by steroid hormones are rare, but various cell culture models exist. Studies in different human blood cell culture models suggest that steroid hormones alter gene expression in human blood cells [76–78]. An in vivo study on macaca fascicularis demonstrates that testosterone and the SARM LGD2941 influence the expression of apoptotic and proliferative genes in blood cells [79].

Reiter et al. [80] could monitor gene expression changes in cell culture experiments with human hair follicle dermal papilla cells that were treated with stanozolol. In another in vivo study, they could show that it is possible to extract RNA out of hair follicle cells and that gene expression in these cells is also influenced by steroid hormones [81]. As shown, blood and hair roots represent promising tissues to find gene expression biomarkers with potential to develop a non-invasive screening method based on gene expression patterns.

### 4.2. Proteomics

The term proteomics describes the study of the proteome which is the actual content of all proteins present in a cell, tissue or organism at a specific physiological stage or as a reaction to a certain treatment.

The use of proteomics for biomarker screening is already common in clinical diagnosis and research. In the diagnosis of different diseases or physiological states blood protein biomarkers are routinely used. Troponins for example are indicators for heart attack, alkaline phosphatase for biliary problems and human chorionic gonadotropin (hCG) is the ultimative marker for early pregnancy [82,83].

In cancer research, as malignant transformation and clonal proliferation of altered cells go in line with alterations in protein expression, proteomics can be used for diagnostic purpose and early detection of cancer [83,84].

Advanced methods for proteomic investigations include two-dimensional gel electrophoresis (2D-gel), mass spectrometry and protein microarrays [38,41] which can be used for biomarker research. These methods are suitable to screen for all multisided changes in protein expression that are caused by a changed physiological status or induced by a specific treatment e.g. by anabolic steroid hormones. This way of biomarker screening can be named as "de novo" approach [82,85] with the advantage, that numerous proteomic changes, also those that are so far unknown can be evaluated. Another way of screening for biomarkers is the evaluation of candidate proteins by screening the actual literature or by regarding physiological effects that are present at a specific physiological state or are induced by a certain treatment [82,85]. Methods that can be used for this approach are 1-D electrophoresis, RIA (radio immuno assay), ELISA (enzyme linked immuno sorbant assay) or western blot. The advantage of these methods is that only a numerable number of proteins have to be analyzed, immuno assays are more sensitive and the analysis of data sets is well arranged.

Regarding known effects of steroid hormones on protein expression or excretion it could be investigated if the candidate protein approach will be a promising way for developing a potential screening method for the application of anabolic agents.

Very promising proteins for developing a protein biomarker pattern will be IGF-1, IGF-1BP3 and somatotropin (ST). Numerous reports showed these proteins to be increased after the use of anabolic agents in blood plasma of animals and humans [86–93]. Clenbuterol has been shown to down regulate the beta adrenergic receptor and the glucocorticoid receptor in blood cells of veal calves. Dexamethasone also down regulates plasma levels of the glucocorticoid receptor in calves [94]. Different lipoproteins or apolipoproteins are also affected by anabolic hormones in cattle and humans. Hartgens and coworkers [95] could show that androgenic anabolic hormones (AAS) increase plasma protein levels of low density lipoprotein cholesterol and apolipoprotein B and decrease protein levels of high density lipoprotein cholesterol, apolipoprotein A1 and lipoprotein(a) in athletes. An increase of apolipoprotein A1 in plasma of calves by the xenobiotic androgen boldone was demonstrated by Draisci et al. [96]. Propeptide of type III procollagen is known to be a potential marker for the use of anabolic agents in humans [97]. In female calves treated with a combination of oestradiol plus nortestosterone the content of propeptide of type III procollagen is also increased [98]. In the same animal trial it was shown that the combination of nortestosterone plus oestradiol decreases plasma ir-inhibin levels in male calves and that treatment with dexamethasone decreases plasma osteocalcin in veal calves independent of gender [99]. Gardini et al. [100] tried to evaluate protein biomarkers for the treatment of calves with an anabolic combination of estradiol-17β, clenbuterol and dexamethasone by using the combination of 2D-gel and mass spectrometry. They found two regulated proteins in liver tissue (up-regulation of reticulocalbin, down-regulation of adenosine kinase) which could be possible new biomarker candidates for the treatment with this drug combination [100,101].

Apoptotic factors and pro- and anti-inflammatory factors are also promising biomarker candidates because of the known effects of anabolic steroid hormones on apoptosis in different tissues [102–104] and the immune response, in which estrogens show pro-inflammatory and androgens anti-inflammatory effects [105].

Although the proteomic approach is a very promising way to develop a biomarker screening pattern, but up to now very few publications are available in the open literature.

### 4.3. Metabolomics

The metabolome is the collectivity of small-molecule nutrients and metabolites (e.g. metabolic intermediates) in a biological sample. The term metabolomics (also metabonomics) has been established in analogy to transcriptomics and proteomics and describes the study of the metabolome at a certain time point.

Other than transcriptomics and proteomics there is no preferential technique for metabolic investigations so far. In former times changes in the metabolome were measured by detecting single metabolites or degradation products of the certain metabolic pathways in body fluids like urine or blood by chromatograpical or kinetic methods (e.g. Jaffé reaction for the detection of creatinine). The concentration of metabolites like glucose or fatty acids could also be determined by enzymatic methods (e.g. glucose oxidase method) or colorimetric methods using commercially available kits [106,107]. Nowadays due to the technological developments and the availability of hundreds of different standards it is possible to simultaneously measure a great number of substances in one assay to reflect the metabolic status of a certain cell. This metabolic screening method is used for biomarker development mainly in research fields concerning cancer or other diseases using gas or liquid chromatography coupled with mass spectrometry and NMR spectroscopy [108–111]. Although there are few efforts by now, these technologies could also be applied for metabolomic studies in the investigation of hormone function in the organism.

The anabolic effects of β-agonists are mainly due to an increase in muscle protein deposition and a decrease in fat accretion [112]. Metabolites that are involved in these mechanisms could act as potential biomarkers for the use of β-agonists. Creatinine, an indicator for muscle protein synthesis, and Nτ-methylhistidine (MH), an indicator for muscle protein degradation, act as metabolic indicators for protein metabolism [113,114]. Williams et al. [114] found a higher creatinine excretion and a reduction of MH elimination in the urine of animals fed with clenbuterol compared to the control group.

The decrease in body fat due to the application of β-agonists can be explained by an induction of lipolysis and an inhibition of lipogenesis. Higher concentrations of non-esterified fatty acids (NEFA) in the plasma of animals treated with β-agonists occurred in several studies [106,107,115,116]. Not only the plasma concentration of NEFA, but also the fatty acid composition in the plasma was shown to be changed by clenbuterol [107].

Various studies showed an increase in the plasma glucose levels due to enhanced gluconeogenesis and glycogenolysis in the liver and the breakdown of muscle glycogen to supply the energy sources for the formation of muscle protein after the administration of β-agonists [107,115–118]. Natural catecholamines have been shown to exert indirect mechanisms on the release of several hormones, e.g. the inhibition of the insulin release and thereby the insulin-mediated glycolysis and glycogenesis [118]. In contrast, under the influence of synthetic β-agonists an increase in the insulin level could be observed [112,119].

As concurrently an increase in the glucose, lactate and NEFA plasma levels occurs the development of an insulin resistance under the treatment is suggested. The release of energy substrates goes along with an increase in the blood flow to alleviate the transport to the target tissues [106,115,120]. Equally large amounts of lactate occur in the plasma suggesting an increased glycolysis in muscle tissue for the formation of ATP as energy source for the development of muscle mass [112,116,117].

The effects of β-agonists are mainly transient and the initially marked response becomes attenuated due to a lower responsiveness and a down-regulation of β-adrenergic receptors [120,121].

There have also been efforts to detect perturbations in the metabolic profile after the administration of steroid hormones to reveal the illicit application as growth promoters. Blood metabolites like creatinine or creatine kinase, which are associated with muscle function, or the plasma urea levels, which are known to be an early indicator for anabolic effects in cattle, could serve as potential biomarkers for treatment screening. Mooney et al. [98] measured this metabolites by UV based enzymatic and kinetic methods and showed a significant increase in the plasma urea levels under the influence of estradiol-17β benzoate plus nortestosterone decanoate, but no alteration in the creatinine levels or the creatine kinase activity compared to the control animals.

Cunningham et al. [122] investigated different blood metabolites of ruminants treated with anabolic steroids by standard blood chemistry analysis to investigate if these parameters could be used in a screening test to detect illegal use of growth promoting hormones. Herein no significant change in the urea levels between treated and untreated animals, but a significant increase in the creatine levels of heifers treated with nortestosterone decanoate and steers treated with estradiol benzoate was demonstrated. Also substance specific effects on the billirubin levels were shown, with levels being increased in steers and decreased in heifers. As these effects just occur on certain days and not over the complete course of the study period these parameter could not serve as marker for anabolic treatment [122].

In human sports the investigation of the steroid profile is used as a versatile screening tool for routine doping control. The steroid pattern in urine shows distinct ratios of several endogenously synthesized steroid hormones due to the natural excretion. These ratios can be perturbated by increasing or decreasing certain steroid concentrations during the administration of exogenous anabolic compounds [123]. As these ratios can also be altered by natural reasons like the belonging to different ethnic groups this measurement is not sufficient to prove a doping suspicion. GC-C-IRMS can be used to reveal the origin of the applied substance as exogenously applied and endogenously synthesized steroids vary in the ratio of the carbon isotopes $^{12}C$ and $^{13}C$. Pharmaceutical steroids show lower amounts of $^{13}C$, as they are not synthesized de novo, but derived from plant materials [124].

However compared to the β-adrenergic agonists few is known about the metabolic effects of steroid hormones and more targeted investigation have to be done to make a statement on changes in body fluids.

## 5. Bioinformatics

Regardless of which *omic* technology is used for biomarker research, bioinformatical tools are necessary to extract the needed information from the resulting data set.

There are very few examples of unequivocal evidence given by a single biomarker like the trophoblast marker hCG being only present in early pregnancy. In biomarker research the scientist gets a pattern of biomarkers with multiple factors being influenced quantitatively by the drug or the specific physiological stage. The most important question is how do deal with a huge data set to extract, interpret and visualize the intended information. To transform the high-dimensional data into a reduced subspace for representing data in far fewer dimensions, methods for dimensionality reduction are needed [125]. In combination with pattern recognition technologies the identification and visualization of the desired information is approached.

A simple method to classify samples by genomic, proteomic or metabolomic expression patterns is two or three dimensional scatter plot [61]. Using this method only two or three transcripts, proteins or metabolites can be included [60,61]. If more components should be taken into account, multivariate analysis methods are required. Principle components analysis (PCA) reduces multidimensional data sets to lower dimensions called "principle components" [60,61,126]. Each analyzed sample will be visualized by one spot that results from diminishing all collected data of the specific sample to two principle components and so each analyzed sample will be represented by one spot. Employing this method for growth promoter treatment screening will ideally result in a graph with two groups of spots. One group representing the untreated controls and the other group representing the treated individuals. PCA was effectively used by Riedmaier et al. [71] to identify potential gene expression biomarker patterns for anabolic treatment in bovine blood.

To combine the results of two *omic* technologies canonical correlation analysis (CCA) can be used. This method summarizes the relationship between two sets of variables and shows what is common amongst the two sets [126]. To combine the results of two or more *omic* technologies linear or multiple discriminant analysis (LDA or MDA) can be used [126]. Based on a set of measurable fea-

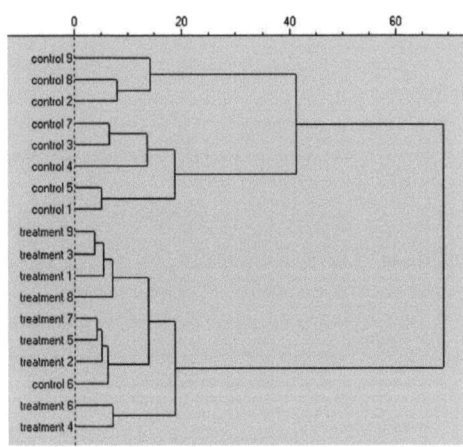

Fig. 4. Cluster dendogram of a qRT-PCR experiment with hormone treated and untreated heifers. Quantitative RT-PCR expression results of vaginal epithelial cells of untreated (control) and treated (treatment) heifers were clustered using GenEx version 4.3.6 Software (MultiD Analyses AB, Gothenburg, Sweden).

tures, these methods classify objects into groups. Screening for the use of anabolic hormones by CCA, LDA or MDA should result in a picture similar to those of PCA, where the treated individuals are separated from the group of the untreated controls.

Another method for visualizing treatment patterns based on multivariate data is hierarchical cluster analysis. The hierarchical order is represented by a tree dendogram in which related samples are more closely together than samples that are more different [60,126]. Used in anabolic treatment screening hierarchical cluster analysis should result in a tree where the treated or the untreated samples respectively are close together and the group of treated samples is separated from the group of untreated samples. Fig. 4 shows a dendogram of qRT-PCR data received from vaginal epithelial cells of heifers, treated with anabolic hormones (unpublished data). The treated ($n$ = 9) and untreated ($n$ = 9) individuals are close together.

Regardless of which biostatistical method will be employed for treatment screening, it is always necessary to have a high number of untreated controls serving as basis for physiological normal individuals. High biological variance between each individual are caused by genetically diversity or environmental conditions [127]. To deal with these differences between various individuals a high number of control samples representing the investigated group of animals of humans is needed.

In summary with advanced biostatistical method marking and reliable classification of treated animals is possible.

## 6. Conclusions

The use of *omic* technologies will be a promising way to develop new screening methods for the detection of the misuse of anabolic steroids and β-agonists based on the physiological changes caused by these substances. Very sensitive methods, like quantitative RT-PCR and mass spectrometry allow the quantification of very small changes in gene expression, protein expression or in the presence of metabolites. With the help of biostatistical tools it is possible to extract the quested information from the resulting data sets.

The discovery of newly designed substances, new modes of drug misuse or other kinds of manipulation in animal husbandry or sports – like erythropoietin, blood or gene doping – will be a future challenge to *omic* techniques.

## References

[1] H.H.D. Meyer, APMIS 109 (2001) 1.
[2] M. Hau, Bioessays 29 (2007) 2.
[3] J.E. Morley, R.N. Baumgartner, R. Roubenoff, J. Mayer, K.S. Nair, J. Lab. Clin. Med. 137 (2001) 4.
[4] J.E. Morley, Nutrition 17 (2001) 7.
[5] J.E. Morley, M.J. Kim, M.T. Haren, Rev. Endocr. Metab. Disord. 6 (2005) 2.
[6] T.J. Marcell, J. Gerontol. A Biol. Sci. Med. Sci. 58 (2003) 10.
[7] F.J. Lopez, Curr. Opin. Chem. Biol. 4 (2000) 4.
[8] M.R. Deschenes, Sports Med. 34 (2004) 12.
[9] T.J. Doherty, J. Appl. Physiol. 95 (2003) 4.
[10] A. Negro-Vilar, J. Clin. Endocrinol. Metab. 84 (1999) 10.
[11] C. Saudan, N. Baume, N. Robinson, L. Avois, P. Mangin, M. Saugy, Br. J. Sports Med. 40 (Suppl. 1) (2006).
[12] G. Mazzanti, C. Daniele, G. Boatto, G. Manca, G. Brambilla, A. Loizzo, Toxicology 187 (2003) 2.
[13] A.W. Bell, D.E. Bauman, D.H. Beermann, R.J. Harrell, J. Nutr. 128 (2 Suppl.) (1998).
[14] A. Chwalibog, K. Jensen, G. Thorbek, Arch. Tierernahr. 49 (1996) 2.
[15] L.O. Fiems, C.V. Boucque, B.G. Cottyn, Arch. Tierernahr. 45 (1993) 2.
[16] K. Reichel, C. Rehfeldt, R. Weikard, R. Schadereit, L. Krawielitzki, Arch. Tierernahr. 45 (1993) 3.
[17] H.A. Kuiper, M.Y. Noordam, M.M. Dooren-Flipsen, R. Schilt, A.H. Roos, J. Anim. Sci. 76 (1998) 1.
[18] H.H.D. Meyer, L.M. Rinke, J. Anim. Sci. 69 (1991) 11.
[19] H.J. Mersmann, J. Anim. Sci. 76 (1998) 1.
[20] I.G. Lange, A. Daxenberger, H.H.D. Meyer, APMIS 109 (2001) 1.
[21] C. Moran, J.F. Quirke, D.J. Prendiville, S. Bourke, J.F. Roche, J. Anim. Sci. 69 (1991) 11.
[22] A. Daxenberger, D. Ibarreta, H.H.D. Meyer, Hum. Reprod. Update 7 (2001) 3.
[23] S.H. Swan, F. Liu, J.W. Overstreet, C. Brazil, N.E. Skakkebaek, Hum. Reprod. 22 (2007) 6.
[24] D. Maume, Y. Deceuninck, K. Pouponneau, A. Paris, B. Le Bizec, F. Andre, APMIS 109 (2001) 1.
[25] R. Schilt, M.J. Groot, P.L. Berende, V. Ramazza, J.S. Ossenkoppele, W. Haasnoot, E.O. Van Bennekom, L. Brouwer, H. Hooijerink, Analyst 123 (1998) 12.
[26] R.W. Stephany, APMIS Suppl. (2001) 103.
[27] H.K. Beckett, D.A. Cowan, Br. J. Sports Med. 12 (1978) 4.
[28] D.H. Catlin, T.H. Murray, JAMA 276 (1996) 3.
[29] L.E. Grivetti, E.A. Applegate, J. Nutr. 127 (5 Suppl.) (1997).
[30] H.M. Prendergast, T. Bannen, T.B. Erickson, K.R. Honore, Vet. Hum. Toxicol. 45 (2003) 2.
[31] W.W. Franke, B. Berendonk, Clin. Chem. 43 (1997) 7.
[32] F. Sjoqvist, M. Garle, A. Rane, Lancet 371 (2008) 9627.
[33] W. Schanzer, M. Thevis, Med. Klin. (Munich) 102 (2007) 8.
[34] H.H.D. Meyer, S. Hoffmann, Food Addit. Contam. 4 (1987) 2.
[35] H.H.D. Meyer, L. Rinke, I. Dursch, J. Chromatogr. 564 (1991) 2.
[36] M.L. Scippo, G. Degand, A. Duyckaerts, G. Maghuin-Rogister, P. Delahaut, Analyst 119 (1994) 12.
[37] L.A. van Ginkel, J. Chromatogr. 564 (1991) 2.
[38] K.F. Becker, V. Metzger, S. Hipp, H. Hofler, Curr. Med. Chem. 13 (2006) 15.
[39] J.A. Ludwig, J.N. Weinstein, Nat. Rev. Cancer 5 (2005) 11.
[40] S.S. Ilyin, S.M. Belkowski, C.R. Plata-Salaman, Trends Biotechnol. 22 (2004) 8.
[41] X. Zhang, L. Li, D. Wei, Y. Yap, F. Chen, Trends Biotechnol. 25 (2007) 4.
[42] M. Beato, S. Chavez, M. Truss, Steroids 61 (1996) 4.
[43] D.P. Edwards, Annu. Rev. Physiol. 67 (2005).
[44] D. Picard, V. Kumar, P. Chambon, K.R. Yamamoto, Cell Regul. 1 (1990) 3.
[45] M. Rexin, W. Busch, B. Segnitz, U. Gehring, J. Biol. Chem. 267 (1992) 14.
[46] P. Rehberger, M. Rexin, U. Gehring, Proc. Natl. Acad. Sci. U.S.A. 89 (1992) 17.
[47] B. Segnitz, U. Gehring, Proc. Natl. Acad. Sci. U.S.A. 92 (1995) 6.
[48] M. Beato, Cell 56 (1989) 3.
[49] A. Griekspoor, W. Zwart, J. Neefjes, R. Michalides, Nucl. Recept. Signal. 5 (2007).
[50] N.L. Weigel, N.L. Moore, Nucl. Recept. Signal. 5 (2007).
[51] A. Philips, D. Chalbos, H. Rochefort, J. Biol. Chem. 268 (1993) 19.
[52] P. Webb, G.N. Lopez, R.M. Uht, P.J. Kushner, Mol. Endocrinol. 9 (1995) 4.
[53] B. Stein, M.X. Yang, Mol. Cell Biol. 15 (1995) 9.
[54] N.H. Ing, Biol. Reprod. 72 (2005) 6.
[55] G.G. Kuiper, B. Carlsson, K. Grandien, E. Enmark, J. Haggblad, S. Nilsson, J.A. Gustafsson, Endocrinology 138 (1997) 3.
[56] M.W. Pfaffl, A. Daxenberger, M. Hageleit, H.H.D. Meyer, J. Vet. Med. A Physiol. Pathol. Clin. Med. 49 (2002) 2.
[57] M.W. Pfaffl, I.G. Lange, H.H.D. Meyer, J. Steroid Biochem. Mol. Biol. 84 (2003) 2.
[58] D. Strosberg, Pathol. Biol. (Paris) 40 (1992) 8.
[59] S.B. Liggett, J.R. Raymond, Baillieres Clin. Endocrinol. Metab. 7 (1993) 2.
[60] J. Logan, J. Edwards, N. Saunders, Applied and Functional Genomics, Health Protection Agency, Real-Time PCR: Current Technology and Applications. Caister Academic Press, London, GB, 2009, p. 5.
[61] M. Kubista, J.M. Andrade, M. Bengtsson, A. Forootan, J. Jonak, K. Lind, R. Sindelka, B. Sjoback, B. Sjogreen, L. Strombom, A. Stahlberg, N. Zoric, Mol. Aspects Med. 27 (2006) 2.
[62] A. Abdullah-Sayani, J.M. Bueno-de-Mesquita, M.J. van de Vijver, Nat. Clin. Prac Oncol. 3 (2006) 9.
[63] G.S. Ginsburg, S.B. Haga, Expert. Rev. Mol. Diagn. 6 (2006) 2.
[64] J. Quackenbush, New Engl. J. Med. 354 (2006) 23.
[65] A.K. Sandvik, B.K. Alsberg, K.G. Norsett, F. Yadetie, H.L. Waldum, A. Laegreid, Clin. Chim. Acta 363 (2006) 1.
[66] A.V. Tinker, A. Boussioutas, D.D. Bowtell, Cancer Cell 9 (2006) 5.
[67] M. Reiter, V.M. Walf, A. Christians, M.W. Pfaffl, H.H.D. Meyer, Anal. Chim. Act 586 (2007) 1.
[68] B.J. Johnson, M.E. White, M.R. Hathaway, C.J. Christians, W.R. Dayton, J. Anim. Sci. 76 (1998) 2.
[69] M.E. White, B.J. Johnson, M.R. Hathaway, W.R. Dayton, J. Anim. Sci. 81 (2003) 4.
[70] L. Toffolatti, G.L. Rosa, T. Patarnello, C. Romualdi, R. Merlanti, C. Montesissa, Poppi, M. Castagnaro, L. Bargelloni, Domest. Anim. Endocrinol. 30 (2006) 1.
[71] I. Riedmaier, A. Tichopad, M. Reiter, M.W. Pfaffl, H.H.D. Meyer, Anal. Chim. Act doi:10.1016/j.aca.2009.02.014.
[72] L.C. Chang, S.A. Madsen, T. Toelboell, P.S. Weber, J.L. Burton, J. Endocrinol. 18 (2004) 3.
[73] M. Cantiello, M. Carletti, F.T. Cannizzo, C. Nebbia, C. Bellino, S. Pie, I.P. Oswald E. Bollo, M. Dacasto, Toxicology 242 (2007) 1.
[74] S. Sato, N. Nomura, F. Kawano, J. Tanihata, K. Tachiyashiki, K. Imaizumi, J. Pha macol. Sci. 107 (2008) 4.
[75] D.H. Beermann, J. Anim. Sci. 80 (2002).
[76] S. Capellino, V. Villaggio, P. Montagna, A. Sulli, C. Craviotto, M. Cutolo, Reuma tismo 57 (2005) 3.
[77] N. Kanda, S. Watanabe, J. Invest. Dermatol. 118 (2002) 3.
[78] P.R. Kramer, S.F. Kramer, G. Guan, Arthritis Rheum. 50 (2004) 6.
[79] I. Riedmaier, A. Tichopad, M. Reiter, M.W. Pfaffl, H.H.D. Meyer, J. Steroi Biochem. Mol. Biol. (2008), doi:10.1016/j.jsbmb.2009.01.019.
[80] M. Reiter, M.W. Pfaffl, M. Schoenfelder, H.H.D. Meyer, Biomarker Insights (2009).
[81] M. Reiter, S. Lüderwald, M.W. Pfaffl, H.H.D. Meyer, Doping J., dj052008-01.
[82] A.G. Paulovich, J.R. Whiteaker, A.N. Hoofnagle, P. Wang, Proteomics Clin. App 2 (2008).
[83] N. Ramachandran, S. Srivastava, J. LaBaer, Proteomics Clin. Appl. 2 (2008).
[84] W.C. Cho, Mol. Cancer 6 (2007).
[85] P. Matt, Z. Fu, Q. Fu, J.E. Van Eyk, Physiol. Genom. 33 (2008) 1.
[86] B.J. Johnson, M.R. Hathaway, P.T. Anderson, J.C. Meiske, W.R. Dayton, J. Anim Sci. 74 (1996) 2.
[87] P.J. Buttery, J.M. Dawson, Proc. Nutr. Soc. 49 (1990) 3.
[88] D.D. Hongerholt, B.A. Crooker, J.E. Wheaton, K.M. Carlson, D.M. Jorgenson, Anim. Sci. 70 (1992) 5.
[89] E.J. Duschek, L.J. Gooren, C. Netelenbos, Maturitas 51 (2005) 3.
[90] M. Alen, P. Rahkila, M. Reinila, R. Vihko, Am. J. Sports Med. 15 (1987) 4.
[91] J.O. Jorgensen, J.J. Christensen, M. Krag, S. Fisker, P. Ovesen, J.S. Christiansen Horm. Res. 62 (Suppl. 1) (2004).
[92] R.D. Cluckman, R.H. Breier, S.R. Davis, J. Dairy Sci. 70 (1987) 2.
[93] W.J. Enright, L.T. Chapin, W.M. Moseley, S.A. Zinn, M.B. Kamdar, L.F. Krabill H.A. Tucker, J. Endocrinol. 122 (1989) 3.
[94] R. Odore, P. Badino, S. Pagliasso, C. Nebbia, B. Cuniberti, R. Barbero, G. Re, Vet. Pharmacol. Ther. 29 (2006) 2.
[95] F. Hartgens, G. Rietjens, H.A. Keizer, H. Kuipers, B.H. Wolffenbuttel, Br. J. Sport Med. 38 (2004) 3.
[96] R. Draisci, C. Montesissa, B. Santamaria, C. D'Ambrosio, G. Ferretti, R. Merlant C. Ferranti, M. De Liguoro, C. Cartoni, E. Pistarino, L. Ferrara, M. Tiso, A. Scalon M.E. Cosulich, Proteomics 7 (2007) 17.
[97] R. Abellan, R. Ventura, S. Pichini, J.A. Pascual, R. Pacifici, S. Di Carlo, A. Bacos J. Segura, P. Zuccaro, Clin. Chem. Lab. Med. 43 (2005) 1.
[98] M.H. Mooney, C. Situ, G. Cacciatore, T. Hutchinson, C. Elliott, A.A. Bergwerf Biomarkers 13 (2008) 3.
[99] G. Cacciatore, S.W.F. Eisenberg, C. Situ, M.H. Mooney, P. Delahaut, S. Klaren beek, A.C. Huet, A.A. Bergwerff, C.T. Elliott, Anal. Chim. Acta (2008 doi:10.1016/j.aca.2008.11.027.
[100] G. Gardini, P. Del Boccio, S. Colombatto, G. Testore, D. Corpillo, C. Di Ilio, A Urbani, C. Nebbia, Proteomics 6 (2006) 9.
[101] C. Nebbia, D.L. Della, M. Carletti, A. Balbo, G. Barbarino, G. Gardini, J. Vet Pharmacol. Ther. 31 (2008) 3.
[102] M. Cutolo, S. Capellino, P. Montagna, P. Ghiorzo, A. Sulli, B. Villaggio, Arthriti Res. Therapy 7 (2005) 5.
[103] R. Boland, A. Vasconsuelo, L. Milanesi, A.C. Ronda, A.R. de Boland, Steroids 7: (2008) 9.
[104] R.W. McMurray, S. Suwannaroj, K. Ndebele, J.K. Jenkins, Pathobiology 6! (2001) 1.
[105] D. Janele, T. Lang, S. Capellino, M. Cutolo, J.A. Da Silva, R.H. Straub, Ann. N Acad. Sci. 1069 (2006).
[106] T.M. Byrem, D.H. Beermann, T.F. Robinson, J. Anim. Sci. 76 (1998) 4.
[107] B. Stoffel, H.H.D. Meyer, J. Anim. Sci. 71 (1993) 7.
[108] W.M. Claudino, A. Quattrone, L. Biganzoli, M. Pestrin, I. Bertini, L.A. Di, J. Clin Oncol. 25 (2007) 19.
[109] G.D. Lewis, A. Asnani, R.E. Gerszten, J. Am. Coll. Cardiol. 52 (2008) 2.
[110] S. Cubbon, T. Bradbury, J. Wilson, J. Thomas-Oates, Anal. Chem. 79 (2007) 23.
[111] M.E. Dumas, C. Canlet, J. Vercauteren, F. Andre, A. Paris, J. Proteome Res. (2005) 5.
[112] J.W. Blum, N. Flueckiger, Eur. J. Pharmacol. 151 (1988) 2.

[113] E.G. Afting, W. Bernhardt, R.W. Janzen, H.J. Rothig, Biochem. J. 200 (1981) 2.
[114] P.E. Williams, L. Pagliani, G.M. Innes, K. Pennie, C.I. Harris, P. Garthwaite, Br. J. Nutr. 57 (1987) 3.
[115] J.H. Eisemann, G.B. Huntington, C.L. Ferrell, J. Anim. Sci. 66 (1988) 2.
[116] O. Adeola, B.W. McBride, L.G. Young, J. Nutr. 122 (1992) 6.
[117] J.B. Li, L.S. Jefferson, Am. J. Physiol. 232 (1977) 2.
[118] E. Barth, G. Albuszies, K. Baumgart, M. Matejovic, U. Wachter, J. Vogt, P. Radermacher, E. Calzia, Crit. Care Med. 35 (9 Suppl.) (2007).
[119] J.K. Zimmerli, J.W. Blum, J. Anim. Physiol. Anim. Nutr. 63 (1990).
[120] J.K. Eadara, R.H. Dalrymple, R.L. DeLay, C.A. Ricks, D.R. Romsos, Metabolism 38 (1989) 9.
[121] W.P. Hausdorff, M.G. Caron, R.J. Lefkowitz, FASEB J. 4 (1990) 11.
[122] R.T. Cunningham, M.H. Mooney, X.L. Xia, S. Crooks, D. Matthews, M. O'Keeffe, K. Li, C.T. Elliott, Anal. Chem. (2009).
[123] U. Mareck, H. Geyer, G. Opfermann, M. Thevis, W. Schanzer, J. Mass Spectrom. 43 (2008) 7.
[124] T. Piper, U. Mareck, H. Geyer, U. Flenker, M. Thevis, P. Platen, W. Schanzer, Rapid Commun. Mass Spectrom. 22 (2008) 14.
[125] G. Lee, C. Rodriguez, A. Madabhushi, IEEE/ACM. Trans. Comput. Biol. Bioinform. 5 (2008) 3.
[126] J. Beyene, D. Tritchler, S.B. Bull, K.C. Cartier, G. Jonasdottir, A.T. Kraja, N. Li, N.L. Nock, E. Parkhomenko, J.S. Rao, C.M. Stein, R. Sutradhar, S. Waaijenborg, K.S. Wang, Y. Wang, P. Wolkow, Genet. Epidemiol. 31 (Suppl. 1) (2007).
[127] C.C. Pritchard, L. Hsu, J. Delrow, P.S. Nelson, Proc. Natl. Acad. Sci. U.S.A. 98 (2001) 23.

# Appendix II

# Journal of Steroid Biochemistry and Molecular Biology

journal homepage: www.elsevier.com/locate/jsbmb

## Influence of testosterone and a novel SARM on gene expression in whole blood of *Macaca fascicularis*☆

Irmgard Riedmaier*, Ales Tichopad, Martina Reiter, Michael W. Pfaffl, Heinrich H.D. Meyer

*Physiology Weihenstephan, Technische Universität München, Weihenstephaner Berg 3, 85354 Freising, Germany*

### ARTICLE INFO

*Article history:*
Received 5 November 2008
Received in revised form 26 January 2009
Accepted 28 January 2009

*Keywords:*
Testosterone
SARM
Biomarker
Gene expression
Real-time qRT-PCR

### ABSTRACT

Anabolic hormones, including testosterone, have been suggested as a therapy for aging-related conditions, such as osteoporosis and sarcopenia. These therapies are sometimes associated with severe androgenic side effects. A promising alternative to testosterone replacement therapy are selective androgen receptor modulators (SARMs). SARMs have the potential to mimic the desirable central and peripheral androgenic anabolic effects of testosterone without having its side effects.

In this study we evaluated the effects of LGD2941, in comparison to testosterone, on mRNA expression of selected target genes in whole blood in an non-human model. The regulated genes can act as potential blood biomarker candidates in future studies with AR ligands.

Cynomolgus monkeys (*Macaca fascicularis*) were treated either with testosterone or LGD2941 for 90 days in order to compare their effects on mRNA expression in blood. Blood samples were taken before SARM application, on day 16 and on day 90 of treatment.

Gene expression of 37 candidate genes was measured using quantitative real-time RT-PCR (qRT-PCR) technology.

Our study shows that both testosterone and LGD2941 influence mRNA expression of 6 selected genes out of 37 in whole blood. The apoptosis regulators CD30L, Fas, TNFR1 and TNFR2 and the interleukins IL-12B and IL-15 showed significant changes in gene expression between control and the treatment groups and represent potential biomarkers for androgen receptor ligands in whole blood.

© 2009 Elsevier Ltd. All rights reserved.

## 1. Introduction

Over the last decades the proportion of elderly people in the population has increased [1]. This is the reason why the incidence of age-related conditions like sarcopenia and osteoporosis is rising and becoming one of the major topics in health care. Sarcopenia is the loss of muscle mass during the aging process that may lead to frailty [2–5]. Sarcopenia is commonly associated with osteoporosis, which is the age-related loss of bone mineral density. The combination of sarcopenia and osteoporosis results in a high incidence of bone fractures relating to accidental falls, which is a significant cause of morbidity and mortality in the elderly population.

Both conditions are associated with a decrease in the endogen production of anabolic hormones, including testosterone [4]. Testosterone treatment has been proposed as a therapy for osteoporosis and frailty in both men and women [6,7]. However, the androgen therapies available today are associated with androgenic side effects, such as skin virilization in women and prostate hypertrophy in men [8–10].

A promising alternative for testosterone replacement therapy is the development of selective androgen receptor modulators (SARMs) [6]. SARMs are synthetic molecules that bind to the androgen receptor exhibiting tissue-selective effects. An "ideal" SARM is an orally active compound that provides an increase in muscle mass and strength and has an anabolic effect on bone density without inducing undesirable androgenic side effects [6]. LGD2941 is a novel non-steroidal, orally active SARM, which has shown potent anabolic activity on bone and muscle in rats and monkeys, but reduced effects on the prostate [7].

It is already known that androgens cause changes in the biochemical pathways of different organs and tissues. Specific enzymes, receptors and cytokines can be activated or suppressed on the cellular mRNA expression level. Using appropriate specific and sensitive quantification methods, like quantitative real-time RT-PCR, such mRNA expression changes are measurable.

The aim of this study was to evaluate the effects of LGD2941, in comparison to testosterone, on mRNA expression of selected target genes in whole blood samples. Whole blood is chosen because samples can easily been taken from the living organism. Furthermore there is evidence in the literature that androgens affect gene

---

☆ The poster version of this manuscript was presented at the Congress in Seefeld, Tirol 2008.
* Corresponding author. Fax: +49 8161 714204.
E-mail address: irmgard.riedmaier@wzw.tum.de (I. Riedmaier).

0960-0760/$ – see front matter © 2009 Elsevier Ltd. All rights reserved.
doi:10.1016/j.jsbmb.2009.01.019

expression of the different blood cells. The regulated genes have the potential to act as blood biomarkers in future studies with AR ligands.

## 2. Materials and methods

### 2.1. Animal experiment

24 male cynomolgus monkeys (*Macaca fascicularis*) were separated to four groups of six animals each. All animals were 5–6 years old, skeletally mature and had an average body weight of 6± kg. The treatments were group 1 (control or oral vehicle group), group 2 (reference group, testosterone group) 3.0 mg/kg Testosteronenanthate as Testoviron®-depot-250 (Schering, Berlin, Germany), dosed biweekly by intramuscular injection, group 3 (intermediate concentration group, SARM1) 1 mg/kg SARM LGD2941 daily and group 4 (high concentration group, SARM10) 10 mg/kg SARM LGD2941 daily. The oral vehicle control and the SARM were dosed once daily for 90 days.

Whole blood samples were taken at three time points. Predose samples were taken after study start without prior treatment. Further samples were taken at day 16 and day 90 of treatment. Duplicate blood samples (2.5 mL each) were transferred into PAXgene blood RNA tubes (BD, Heidelberg, Germany) gently shaken, incubated at room temperature for two hours and stored at −20 °C.

The animal attendance and blood sampling were done by Covance Laboratories GmbH (Münster, Germany) and was conducted with permission from the local veterinary authorities and in accordance with accepted standards of Humane Animal Care.

### 2.2. RNA preparation and qRT-PCR

RNA from blood samples was extracted using the PAXgene Blood RNA Kit (Qiagen, Hilden, Germany) according to the manufacturer's instructions.

To quantify the amount of total RNA extracted, optical density (OD) was measured with the Biophotometer (Eppendorf Biophotometer, Hamburg, Germany) for each sample. RNA purity was calculated with the $OD_{260/280}$ ratio.

RNA integrity and quality control was performed via capillary electrophoresis in the Bioanalyzer 2100 (Agilent Technology, Palo Alto, USA). Eukaryotic total RNA Nano Assay (Agilent Technology) was taken for sample analysis and the RNA Integrity Number (RIN) served as RNA quality parameter. Agilent Bioanalyzer 2100 calculated the RIN value based on a numbering system from 1 to 10 (1 being the most degraded profile, 10 being the most intact) for all samples. A RIN ≥ 6 should be achieved to assure good results in qRT-PCR [11,12].

Candidate genes were chosen by screening the respective literature for androgen and inflammation-related effects on blood cells. Their expression was investigated using listed primers (Table 1). All primers were designed using published human nucleic acid sequences of GenBank (http://www.ncbi.nlm.nih.gov/entrez/query.fcgi). Primer design and optimization was done with primer design program of MWG Biotech (MWG, Ebersberg, Germany) and primer3 (http://frodo.wi.mit.edu/cgi-bin/primer3/primer3_www.cgi) with regard to primer dimer formation, self-priming formation and a constant primer annealing temperature of 60 °C. Newly designed primers were ordered and synthesized at MWG Biotech (Ebersberg, Germany) or Invitrogen (Karlsruhe, Germany). Primer performance testing was done with six optional untreated samples and a no template control (NTC contains only RNAse free water) for each primer set.

Quantitative real-time RT-PCR was performed using SuperScript III Platinum SYBR Green One-Step qPCR Kit (Invitrogen, Carlsbad, USA) by a standard protocol, recommended by the manufacturer. With the kit the master mix was prepared as follows: for one sample it is 5 µL 2× SYBR Green Reaction Mix, 0.5 µL forward primer (10 pmol/µL), 0.5 µL reverse primer (10 pmol/µL) and 0.2 µL SYBR Green One-Step Enzyme Mix (Invitrogen, Carlsbad, USA). 6.2 µL of the master mix was filled in the special 100 µL tubes and 3.8 µL RNA (concentration 1 ng/µL respectively 10 ng/µL) was added. Tubes were closed, placed into the Rotor-Gene 3000 and Analysis Software v6.0 was started (Corbett Life Science, Sydney, Australia). The following one-step qRT-PCR temperature cycling program was used for all genes: Reverse transcription took place at 55 °C for 10 min. After 5 min of denaturation at 95 °C, 40 cycles of real-time PCR with 3-segment amplification were performed consisting of 15 s at 95 °C for denaturation, 30 s at primer dependent temperature for annealing and 20 s at 68 °C for polymerase elongation. The melting step was then performed with slow heating starting at 60 °C with a rate of 0.5 °C per second up to 95 °C with continuous measurement of fluorescence.

Take off points (Ct) and melting curves were acquired by using the "*Comparative quantitation*" and "*Melting curve*" program of the Rotor-Gene 3000 Analysis software v6.0. Only genes with melting curves showing a single peak and no primer dimers were taken for further data analysis. Samples that showed irregular melting peaks were excluded from the quantification procedure.

### 2.3. Selection of target genes

Candidate genes that might be biomarkers in blood were chosen by screening the respective literature for androgen and inflammation-related effects on blood cells. Androgens are known to down-regulate proliferation of lymphocytes [13,14]. Therefore the different pro- and anti-inflammatory interleukins (IL) IL-1β, IL-2, IL-4, IL-6, IL-10, IL-12B, IL-13 and IL-15 and the growth factors tumor growth factor β (TGF-β), insulin growth factor 1 receptor (IGF-1R) were selected for analysis. It was already shown that testosterone influences the rate of apoptotic blood cells [15–17]. Therefore different apoptosis regulators were chosen for analysis: the TNF receptor superfamily member 6 (Fas), its ligand FasL, tumor necrosis factor receptor (TNFR) 1 and 2, their ligand tumor necrosis factor α (TNF-α), B-cell CLL/lymphoma 2 (BCL2), BCL2-like 1 (BCL-XL), Caspase 3 (Casp 3), Caspase 8 (Casp 8), CD30 Ligand (CD30L), the inflammatory factor nuclear factor of kappa light polypeptide gene enhancer in B-cells 1 (p105) (NFκB) and its inhibitor IκB. To determine if the treatment also has an influence on the amount of the different white blood cells, the expression of the cell specific CD Antigens CD4 (T helper cells), CD8 (cytotoxic T cells), CD11b (granulocytes), CD14 (monocytes), CD20 (B-cells), CD25 (activated T cells) and CD69 were measured. Further leukocyte genes that were measured are androgen receptor (AR), tumor necrosis factor β (TNF-β) and CD27 Ligand (CD27L). As genes expressed in reticulocytes, haemoglobin alpha (α-globin), haemoglobin beta (β-globin) and their transcription factors and stabilization factors transcription factor CP2 (CP2), acid phosphatase 1 (αCP1) and upstream transcription factor 1 (USF-1) were chosen. As reference gene candidates β-Actin and glyceraldehyde-3-phosphate dehydrogenase (GAPDH) were measured, whereas β-Actin and GAPDH were chosen as best reference genes by using GenEx Ver 4.3.3 Software (multiD Analyses AB, Gothenburg, Sweden).

### 2.4. Data analysis and statistics

Statistical description of the expression data as well as statistical tests were produced with SAS v. 9.1.3 for Windows. The raw data were the Ct values obtained from each qPCR sample. Each qRT-PCR sample was associated with a blood sample whereas for each experimental animal two blood samples were analysed. Since the

**Table 1**
List of primer pairs used for qRT-PCR analysis.

| Group | Gene | Primer name | Primer sequence 5′ → 3′ | Product length |
|---|---|---|---|---|
| Reference genes | Ubiquitin C | UBC_for | TGA AGA CTC TGA CTG GTA AGA CC | 128 bp |
| | | UBC_rev | CAT CCA GCA AAG ATC AGC CTC | |
| | Actin-β | ActB_for | AGT CCT GTG GCA TCC ACG AA | 148 bp |
| | | ActB_rev | GCA GTG ATC TCC TTC TGC ATC | |
| | GAPDH | GAPDH_for | GAA GGT GAA GGT CGG AGT CAA | 233 bp |
| | | GAPDH_rev | GCT CCT GGA AGA TGG TGA TG | |
| Interleukins | IL-1β | IL1beta_for | GGA CAG GAT ATG GAG CAA CAA G | 121 bp |
| | | IL1beta_rev | AAC ACG CAG GAC AGG TAC AG | |
| | IL-2 | IL2_for2 | GCA ACT CCT GTC TTG CAT TGC | 165 bp |
| | | IL2_rev2 | CAT CCT GGT GAG TTT GGG ATT C | |
| | IL-4 | IL4_for3 | TTC CCC CTC TGT TCT TCC TG | 121 bp |
| | | IL4_rev3 | GTT GTG TTC TTC TGC TCT GTG AG | |
| | IL-6 | IL6_for5 | AGG AGA CTT GCC TGG TGA AA | 179 bp |
| | | IL6_rev5 | CAG GGG TGG TTA TTG CAT CT | |
| | IL-10 | IL10_for2 | AGC CTTCGTC TGA GAT GAT CCA G | 190 bp |
| | | IL10_rev2 | CAT TCT TCA CCT GCT CCA CG | |
| | IL-12b (p40) | IL12B_for | AAG GAG GCG AGG TTC TAA GC | 213 bp |
| | | IL12B_rev | AAG AGC CTC TGC TGC TTT TGA C | |
| | IL-13 | IL-13_for2 | AAT GGC AGC ATG GTA TGG AGC | 124 bp |
| | | IL-13_rev2 | AGA ATC CGC TCA GCA TCC TC | |
| | IL-15 | IL-15_for2 | TCC AGT GCT ACT TGT GTT TAC TTC | 93 bp |
| | | IL-15_rev2 | TAG GAA GCC CTG CAC TGA AAC | |
| Apoptosis regulators | Fas | FasR_for4 | TTC TGC CAT AAG CCC TGT CC | 174 bp |
| | | FasR_rev4 | CCA CTT CTA AGC CAT GTC CTT C | |
| | Fas ligand | FasL_for2 | GGC CTG TGT CTC CTT GTC AT | 162 bp |
| | | FasL_rev2 | GTG GCC TAT TTG CTT CTC CAA AG | |
| | TNFR1 | TNFR1_for | AGC TGC TCC AAA TGC CGA AAG | 147 bp |
| | | TNFR1_rev | CAG AGG CTG CAA TTG AAG CAC | |
| | TNFR2 | TNFR2_for | TGA CCA GAC AGC TCA GAT GTG | 99 bp |
| | | TNFR2_rev | TCC TCA CAG GAG TCA CAC AC | |
| | bcl-2 | bcl2_for2 | GAG GAT TGT GGC CTT CTT TGA G | 170 bp |
| | | bcl2_rev2 | ACA GTT CCA CAA AGG CAT CCC | |
| | TNF-α | TNFa_for | AGG GAC CTC TCT CTA ATC AGC | 104 bp |
| | | TNFa_rev | CTC AGC TTG AGG GTT TGC TAC | |
| | Caspase 3 | Casp3_for | GAA TTG ATG CGT GAT GTT TC | 198 bp |
| | | Casp3_rev | GCA GGC CTG AAT AAT GAA AAG | |
| | Caspase 8 | Casp8_for | TGG CAC TGA TGG ACA GGA G | 230 bp |
| | | Casp8_rev | GCA GAA AGT CAG CCT CAT CC | |
| | bcl-xl | bcl-xl_for | TAA ACT GGG GTC GCA TTG TG | 145 bp |
| | | bcl-xl_rev | TGG ATC CAA GGC TCT AGG TG | |
| | CD30L | CD30L_for | CAT TCC CAA CTC ACC TGA CAA C | 281 bp |
| | | CD30L_rev | GCT CCA ACT TCA GAT CGA CAG | |
| Growth factors | TGF-β | TGFb_for | TAC TAC GCC AAG GAG GTC AC | 239 bp |
| | | TGFb_rev | AGG TAT CGC CAG GAA TTG TTG C | |
| | IGF-1R | IGF1R_for | CAT TTC ACC TCC ACC ACC AC | 151 bp |
| | | IGF1R_rev | AGG CAT CCT GCC CAT CAT AC | |
| CD antigens | CD4 | CD4_for3 | CTA AGC TCC AGA TGG GCA AG | 154 bp |
| | | CD4_rev3 | TGA GTG GCT CTC ATC ACC AC | |
| | CD8 | CD8_for | GGA CTT CGC CTG TGA TAT CTA C | 112 bp |
| | | CD8_rev | AAA CAC GTC TTC GGT TCC TGT G | |
| | CD11b | CD11b_for | GAG AAC AAC ATG CCC AGA ACC | 246 bp |
| | | CD11b_rev | CGG TCC CAT ATG ACA GTC TG | |
| | CD14 | CD14_for | AGA ACC TTG TGA GCT GGA CG | 115 bp |
| | | CD14_rev | ATG GAT CTC CAC CTC TAC TGC | |
| | CD20 | CD20_for | CAA CTG TGA ACC AGC TAA TCC C | 163 bp |
| | | CD20_rev | CCA TTC ATT CTC AAC GAT GCC AG | |
| | CD25 | CD25_for | ATC AGT GCG TCC AGG GAT AC | 196 bp |
| | | CD25_rev | ACG AGG CAG GAA GTC TCA C | |
| | CD69 | CD69_for2 | TTG GCT ACC AGA GGA AAT GCC | 164 bp |
| | | CD69_rev2 | CAG TCC AAC CCA GTG TTC CT | |
| Transcription factors | NFκB | NFkB_for2 | ATC ATC CAC CTT CAT TCT CAA CTT G | 149 bp |
| | | NFkB_rev2 | ATC CTC CAC CAC ATC TTC CTG | |
| | IκBα | IkappaB_for | AAC AGG AGG TGA TGG TGA ATA AGC TG | 138 bp |
| | | IkappaB_rev | CCT TGT AGA TAT CCG CCT GG | |
| Reticulocyte genes | α-globin | alpha-gl._for | AGA CCT ACT TCC CGC ACT TC | 275 bp |
| | | alpha-gl._rev | CAG AAG CCA GGA ACT TGT CC | |
| | β-globin | beta-gl_for | GTC CAC TCC TGA TGC TGT TAT G | 240 bp |
| | | beta-gl. rev | TGT CAC AGT GCA GCT ACA TC | |
| | αCP1 | aCP1_for | CCA CCC ATG AAC TCA CCA TTC | 160 bp |
| | | aCP1_rev | GCA GAG CCA GTG ATA GTA ACC | |
| | USF1 | USF1_for | AGA TTC AGG AAG GTG CAG TGG | 121 bp |
| | | USF1_rev | CCA TTC TCA GTT CGG AAG ACG | |

Table 1 (Continued)

| Group | Gene | Primer name | Primer sequence 5′ → 3′ | Product length |
|---|---|---|---|---|
| | CP2 | CP2_for3 | TCT TCG TTT ACC ATG CCA TCT ATC | 178 bp |
| | | CP2_rev3 | CAT GCT TCT TCC TGA AAG TTC TG | |
| Other genes | Androgen receptor | AR_for | CCA CTT CCT CCA AGG ACA ATT AC | 126 bp |
| | | AR_rev | TGG ACT CAG ATG CTC CAA CG | |
| | TNFβ | TNFb_for | TGC TCA CCT CAT TGG AGA CC | 149 bp |
| | | TNFb_rev | AGT AGA CGA AGT AGA TGC CAC TG | |
| | CD27L | CD27L_for | ACA GGA CCT CAG CAG GAC | 272 bp |
| | | CD27L_rev | GAG GCA ATG GTA CAA CCT TGG | |

amplification efficiency was not known, the assumption of identical amplification efficiency 100% was made, allowing more simple quantification model.

The Ct values of each gene were averaged by arithmetic mean for each animal. The obtained mean Ct values were then translated to normalized expression quantities using two reference genes in a form of normalization index. The normalization index was calculated as an arithmetic mean of the Ct values of the two reference genes:

$$\text{reference index} = \text{mean}(Ct_{ACTB}, Ct_{GAPDH}) \quad (1)$$

Then, the expression of every target gene was calculated relatively to the expression of the housekeeping gene as:

$$\text{normalized expression} = \frac{2^{\text{reference index}}}{2^{Ct\,\text{target gene}}}, \quad (2)$$

where the 2 represents the 100% amplification efficiency. The normalized expressions of the timepoints 16 and 90 days were then divided with the normalized expressions of the baseline (predose), generating the expression ratio $R$ as:

$$R_{\text{timepoint/baseline}} = \frac{\text{normalized expression}_{\text{timepoint}}}{\text{normalized expression}_{\text{baseline}}} \quad (3)$$

The expression ratio $R$ was then analysed statistically. The Box–Whisker plot was constructed to facilitate visual screening of regulated genes (Figs. 1–3).

The objective of the statistical analysis was to disclose genes with significant regulation between control group and any of the treated groups. It was not intended to perform all treatment-to-treatment tests for all genes in order to avoid statistical type I error (false positive difference). Hence, ANOVA model was calculated on the $\log_2$ transformed $R$ values employing the SAS procedure GLM with contrast sentence defining the control group as the comparison group for all treatment groups, thus adjusting the overall test confidence level to the number of relevant comparisons only. Further adjustment of the overall confidence level with respect to number of investigated genes was not performed. Hence, this study is to be considered as purely explorative whereas significant findings here indicate candidate biomarkers. Tests generating significant ($p < 0.05$) results were reviewed based on descriptive parameters of the compared groups and visually by means of the Box–Whisker plots to disclose possible outliers. As comparable trends were observed between the three treatment groups, no further test were produced.

To disclose multivariate response to the treatment, the method of principal component analysis (PCA) was employed using GenEx v. 4.3.3 (multiD Analyses AB, Göteborg, Sweden). PCA involves a mathematical procedure that transforms a number of variables (here normalized expression values) into a smaller number of uncorrelated variables called principal components. By this the dimensionality of the data is reduced to a number of dimensions that can be plotted in a scatter plot, here two dimensions. The first principal component accounts for as much of the variability in the data as possible, and each succeeding component accounts

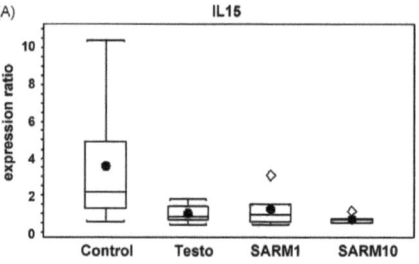
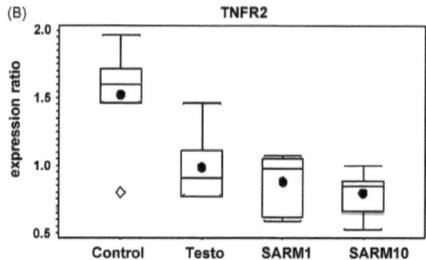

Fig. 1. Significant regulation for IL-15 (A) and TNFR2 (B) between control and treated samples after 16 days of treatment. Box plots show the median, mean (spot) and standard deviations.

for as much of the remaining variability as possible. Normalized expression values of all responding genes were taken as the initial variables and reduced to two principal components only, facilitating thus resolution of treatment clusters in the scatter plot (Fig. 4). Similarly, also each gene was analyzed by PCA taking its response in each sample as the initial variable and plotted in two dimensional scatter plot. This facilitated resolution of co-regulated genes (Fig. 5).

## 3. Results

### 3.1. RNA quality

The mean (±std.dev.) RIN value of the blood samples were 7.5 (±4.8) at predose, 8.5 (±5.0) on day 16 and 7.7 (±4.2) at day 90 indicating a well intact RNA.

### 3.2. Primer testing and gel electrophoresis

Primer pairs of 40 genes were successfully used in qRT-PCR analysis to get single peaks and uniform melting curves, as well as a specific single band in high resolution agarose gel electrophoresis.

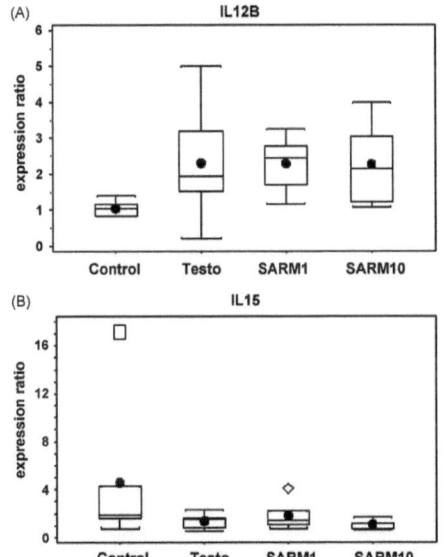

**Fig. 2.** Significant regulation for the proinflammatory interleukins IL-12B (A) and IL-15 (B), between control and treated samples after 90 days of treatment. Box plots show the median, mean (spot) and standard deviations.

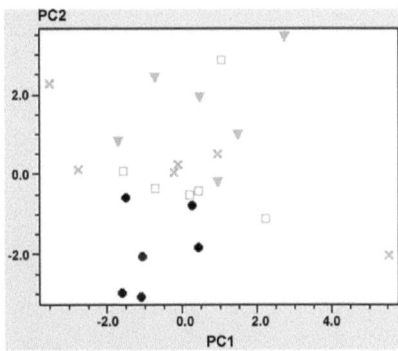

**Fig. 4.** Principal components analysis (PCA) for the six regulated genes IL-12B, IL-15, CD30L, Fas, TNFR1 and TNFR2 in the control group (black dots) the testosterone treated group (grey cross) the low dosed SARM group (grey squares) and the high dosed SARM group (grey triangle).

### 3.3. qRT-PCR results and data analysis

The calculation of the expression ratios (formula (1)) produced non-normally distributed data with frequent extreme values. Some of the extreme values can be outliers and were indicated in the Box–Whisker plot as squares outside the beyond inter quartile range (box). Nonetheless, no exclusion of extreme values/outliers was performed.

Significant down-regulation of gene expression of the treatment groups compared to the control group could be identified for IL-15 ($p = 0.0093$) and TNFR2 ($p < 0.0001$) after 16 days (Fig. 1)

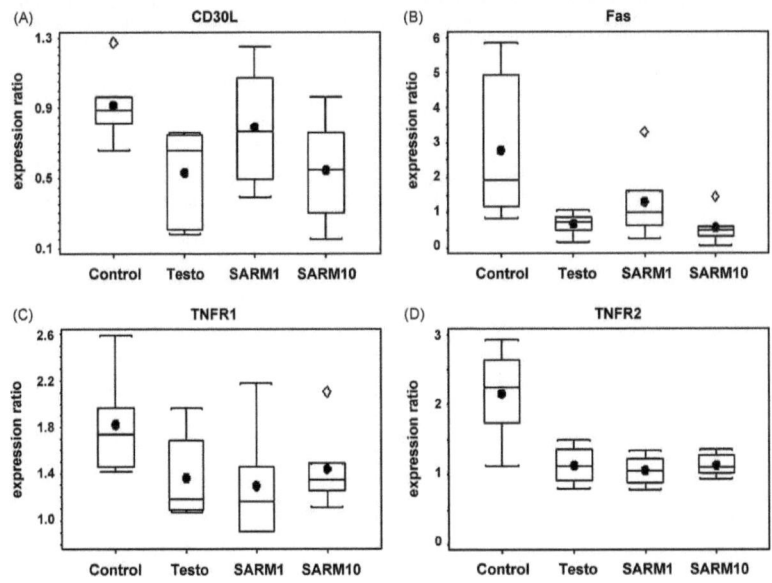

**Fig. 3.** Significant regulation for the apoptosis regulators CD30L (A), Fas (B), TNFR1 (C) and TNFR2 (D) between control and treated samples after 90 days of treatment. Box plots show the median, mean (spot) and standard deviations.

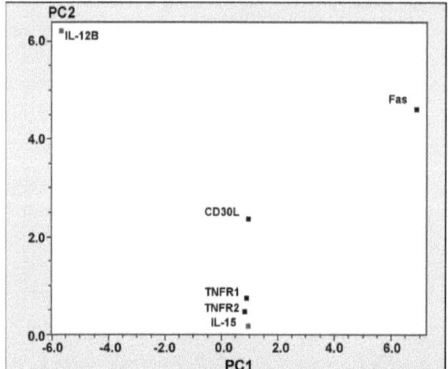

**Fig. 5.** Principle components analysis (PCA) for the regulated genes in all four groups. Grey spots show the interleukins and black spots show the apoptosis regulators.

and for IL-15 ($p = 0.0498$), CD30L ($p = 0.0435$), Fas ($p = 0.0032$), TNFR1 ($p = 0.0308$) and TNFR2 ($p < 0.0001$) after 90 days of treatment. Significant up-regulation of gene expression of the treatment groups compared to the control group could be observed for IL-12B ($p = 0.0240$) after 90 days of treatment (Figs. 2 and 3).

In the control group high variability could be observed compared to the treatment groups as indicated by the Box-Whisker plot. This reflects the natural variability of the non-induced expressing in each studied subject.

Principal components analysis (PCA) is a technique used to reduce multidimensional data sets to lower dimensions for analysis. Fig. 4 was obtained by plotting all samples of the four treatment groups by their two principal components obtained from the six responder genes. Black dots represent samples of the control group, grey crosses show the testosterone group, grey squares represent the SARM1 group and the grey triangles display the SARM10 group. A distinct control group can be seen, showing that there was a multi-transcriptional response to the treatment by any of the three drugs. In addition, the SARM1 neighbors to the control group, creating thus a transition to the Testosteron group and the SARM10 group. In Fig. 5 the six responder genes are clustered. Black dots show apoptosis regulators and grey spots display the interleukins. A distinct cluster of TNF receptors can be resolved.

## 4. Discussion

In this study changes of gene expression in blood cells caused by treatment with LGD2941 or testosterone were evaluated in order to compare the effects of both treatments on gene expression in blood cells. Further aims were the description of physiological effects and the identification of potential biomarkers for the treatment with AR ligands.

The main physiological effect that could be observed in this study is the down-regulation of various apoptotic marker genes in all three treatment groups. This is shown by the significant regulation ($p < 0.05$) of the apoptosis receptors Fas, TNFR1, TNFR2 and the apoptosis ligand CD30L. All regulated apoptosis factors belong either to the TNF Family (CD30L) or to the TNF-Receptor Family (TNFR1, TNFR2, Fas) [18]. It is already known that the death receptor Fas plays a dominant role in the programmed cell death of lymphocytes [18]. When B- and T-cells are activated they get sensitized to Fas mediated apoptosis. On resting peripheral lymphocytes Fas expression is low or even absent. Activation of B- and T-cells results in up-regulation of Fas mRNA [18–23]. Down-regulation of Fas after 90 days of treatment can be a hint to a down-regulating effect on the immune response. The death receptors TNFR1 and TNFR2 activate apoptosis via binding of TNF-α or TNF-β. Binding of the ligand to TNFR1 or TNFR2 can stimulate apoptosis and activate NFκB, whereas in most cases TNFR1 is responsible for these signals [18]. Ligand binding to TNFR2 leads to proliferation of thymocytes [24]. While TNFR2 expression is already regulated after 16 days of treatment, regulation of TNFR1 is only regulated after 90 days of treatment. A reason for this phenomenon could be that the mRNA expression of TNFR2 is inducible whereas expression of TNFR1 is not [24]. CD30L, a member of the TNF ligand superfamily is known to induce apoptosis by binding to its receptor CD30 and is expressed on activated T-cells [25,26]. Down-regulation of CD30L could also be observed after 90 days of treatment.

The down-regulation of these apoptosis regulators suggest that the immune response is suppressed by the treatment with testosterone and the SARM. This observation is consistent with the fact that testosterone has a suppressive effect on the immune system [27–29].

The gene expression of IL-12B - a subunit of IL12 - is up-regulated after 16 days of treatment. The main producers of IL-12 are monocytes, dendritic cells and activated macrophages. It promotes IFN-γ production by CD4 positive T-cells and stimulates proliferation and cytotoxic activity of T-cells and natural killer cells [30]. Gene expression of IL-15 is down-regulated after 16 and 90 days of treatment. It is produced by epithelial cells, fibroblasts, activated monocytes and dendritic cells. It acts as a T-cell activating factor but is not expressed by T-cells themselves [31]. Another important function of IL-15 is the up-regulation of natural killer cell survival and it promotes the production of IFN-γ, GM-CSF and TNF by natural killer cells [32–34].

Regarding the Box-Whisker plots it can be observed that the statistical variance in the control group is higher than in the treatment groups. The reason for this could be the natural variability of the non-induced expression in each studied subject. Suppression of gene expression by an external stimulus like treatment with testosterone or the SARM reduces natural variability of gene expression.

The PCA shows that both drugs show equivalent response and that the treatments differ from the control.

The second aim of this study was to find potential biomarkers for the use of the SARM. If the physiological effects of testosterone and the SARM are compared it became obvious that the SARM is active similar to natural androgens. The regulated genes found in this study can act as first biomarker candidates for the development of a screening pattern in whole blood. To confirm these biomarker candidate genes more studies will be helpful. In primary cell cultures or in further *in vivo* experiments it could be determined if the suggested parameters are independent of age, sex and immune status.

## Acknowledgement

We thank TAP Pharmaceuticals Inc., Lake Forest, USA for supporting this study.

## References

[1] M.C. Walsh, G.R. Hunter, M.B. Livingstone, Sarcopenia in premenopausal and postmenopausal women with osteopenia, osteoporosis and normal bone mineral density, Osteoporos. Int. 17 (1) (2006) 61–67.
[2] J.E. Morley, R.N. Baumgartner, R. Roubenoff, J. Mayer, K.S. Nair, Sarcopenia, J. Lab. Clin. Med. 137 (4) (2001) 231–243.
[3] J.E. Morley, Anorexia, sarcopenia, and aging, Nutrition 17 (7–8) (2001) 660–663.
[4] J.E. Morley, M.J. Kim, M.T. Haren, Frailty and hormones, Rev. Endocr. Metab. Disord. 6 (2) (2005) 101–108.
[5] T.J. Marcell, Sarcopenia: causes, consequences, and preventions, J. Gerontol. A Biol. Sci. Med. Sci. 58 (10) (2003) M911–M916.

[6] A. Negro-Vilar, Selective androgen receptor modulators (SARMs): a novel approach to androgen therapy for the new millennium, J. Clin. Endocrinol. Metab. 84 (10) (1999) 3459–3462.
[7] E. Martinborough, Y. Shen, A. Oeveren, Y.O. Long, T.L. Lau, K.B. Marschke, W.Y. Chang, F.J. Lopez, E.G. Vajda, P.J. Rix, O.H. Viveros, A. Negro-Vilar, L. Zhi, Substituted 6-(1-pyrrolidine)quinolin-2(1H)-ones as novel selective androgen receptor modulators, J. Med. Chem. 50 (21) (2007) 5049–5052.
[8] M.R. Deschenes, Effects of aging on muscle fibre type and size, Sports Med. 34 (12) (2004) 809–824.
[9] T.J. Doherty, Invited review: aging and sarcopenia, J. Appl. Physiol. 95 (4) (2003) 1717–1727.
[10] F.J. Lopez, New approaches to the treatment of osteoporosis, Curr. Opin. Chem. Biol. 4 (4) (2000) 383–393.
[11] S. Fleige, M.W. Pfaffl, RNA integrity and the effect on the real-time qRT-PCR performance, Mol. Aspects Med. 27 (2–3) (2006) 126–139.
[12] A. Schroeder, O. Mueller, S. Stocker, R. Salowsky, M. Leiber, M. Gassmann, S. Lightfoot, W. Menzel, M. Granzow, T. Ragg, The RIN: an RNA integrity number for assigning integrity values to RNA measurements, BMC Mol. Biol. 7 (2006) 3–17.
[13] D. Lehmann, K. Siebold, L.R. Emmons, H. Muller, Androgens inhibit proliferation of human peripheral blood lymphocytes in vitro, Clin. Immunol. Immunopathol. 46 (1) (1988) 122–128.
[14] J.D. Jacobson, M.A. Ansari, Immunomodulatory actions of gonadal steroids may be mediated by gonadotropin-releasing hormone, Endocrinology 145 (1) (2004) 330–336.
[15] S.A. Huber, J. Kupperman, M.K. Newell, Estradiol prevents and testosterone promotes Fas-dependent apoptosis in CD4+ Th2 cells by altering Bcl 2 expression, Lupus 8 (5) (1999) 384–387.
[16] R.W. McMurray, S. Suwannaroj, K. Ndebele, J.K. Jenkins, Differential effects of sex steroids on T and B cells: modulation of cell cycle phase distribution, apoptosis and bcl-2 protein levels, Pathobiology 69 (1) (2001) 44–58.
[17] M. Cutolo, S. Capellino, P. Montagna, P. Ghiorzo, A. Sulli, B. Villaggio, Sex hormone modulation of cell growth and apoptosis of the human monocytic/macrophage cell line, Arthritis Res. Ther. 7 (5) (2005) R1124–R1132.
[18] S. Nagata, Apoptosis by death factor, Cell 88 (3) (1997) 355–365.
[19] T.L. Rothstein, J.K.M. Wang, D.J. Panka, L.C. Foote, Z.H. Wang, B. Stanger, H. Cui, S.T. Ju, A. Marshakrothstein, Protection against Fas-dependent Th1-mediated apoptosis by antigen receptor engagement in B-cells, Nature 374 (6518) (1995) 163–165.
[20] P.H. Krammer, CD95's deadly mission in the immune system, Nature 407 (6805) (2000) 789–795.
[21] C. Klas, K.M. Debatin, R.R. Jonker, P.H. Krammer, Activation interferes with the Apo-1 pathway in mature human T-cells, Int. Immunol. 5 (6) (1993) 625–630.
[22] F. Leithauser, J. Dhein, G. Mechtersheimer, K. Koretz, S. Bruderlein, C. Henne, A. Schmidt, K.M. Debatin, P.H. Krammer, P. Moller, Constitutive and induced expression of Apo-1, a new member of the nerve growth-factor tumor-necrosis-factor receptor superfamily, in normal and neoplastic-cells, Lab. Invest. 69 (4) (1993) 415–429.
[23] H. Chan, D.P. Bartos, L.B. Owen-Schaub, Activation-dependent transcriptional regulation of the human fas promoter requires NF-kappa B p50–p65 recruitment, Mol. Cell. Biol. 19 (3) (1999) 2098–2108.
[24] P. Vandenabeele, W. Declercq, R. Beyaert, W. Fiers, Two tumour necrosis factor receptors: structure and function, Trends Cell Biol. 5 (10) (1995) 392–399.
[25] M.E. Kadin, Regulation of CD30 antigen expression and its potential significance for human disease, Am. J. Pathol. 156 (5) (2000) 1479–1484.
[26] P. Romagnani, F. Annunziato, R. Manetti, C. Mavilia, L. Lasagni, C. Manuelli, G.B. Vannelli, V. Vanini, E. Maggi, C. Pupilli, S. Romagnani, High CD30 ligand expression by epithelial cells and Hassall's corpuscles in the medulla of human thymus, Blood 91 (9) (1998) 3323–3332.
[27] C. Grossman, Possible underlying mechanisms of sexual dimorphism in the immune response, fact and hypothesis, J. Steroid Biochem. 34 (1–6) (1989) 241–251.
[28] V. Morell, Zeroing in on how hormones affect the immune system, Science 269 (5225) (1995) 773–775.
[29] D. Verthelyi, Sex hormones as immunomodulators in health and disease, Int. Immunopharmacol. 1 (6) (2001) 983–993.
[30] M. Del Vecchio, E. Bajetta, S. Canova, M.T. Lotze, A. Wesa, G. Parmiani, A. Anichini, Interleukin-12: biological properties and clinical application, Clin. Cancer Res. 13 (16) (2007) 4677–4685.
[31] K. Liu, M. Catalfamo, Y. Li, P.A. Henkart, N.P. Weng, IL-15 mimics T cell receptor crosslinking in the induction of cellular proliferation, gene expression, and cytotoxicity in CD8+ memory T cells, Proc. Natl. Acad. Sci. U.S.A. 99 (9) (2002) 6192–6197.
[32] Z. Liu, K. Geboes, S. Colpaert, G.R. D'Haens, P. Rutgeerts, J.L. Ceuppens, IL-15 is highly expressed in inflammatory bowel disease and regulates local T cell-dependent cytokine production, J. Immunol. 164 (7) (2000) 3608–3615.
[33] W.E. Carson, M.E. Ross, R.A. Baiocchi, M.J. Marien, N. Boiani, K. Grabstein, M.A. Caligiuri, Endogenous production of interleukin 15 by activated human monocytes is critical for optimal production of interferon-gamma by natural killer cells in vitro, J. Clin. Invest. 96 (6) (1995) 2578–2582.
[34] W.E. Carson, J.G. Giri, M.J. Lindemann, M.L. Linett, M. Ahdieh, R. Paxton, D. Anderson, J. Eisenmann, K. Grabstein, M.A. Caligiuri, Interleukin (IL) 15 is a novel cytokine that activates human natural killer cells via components of the IL-2 receptor, J. Exp. Med. 180 (4) (1994) 1395–1403.

# Appendix III

Contents lists available at ScienceDirect

## Analytica Chimica Acta

journal homepage: www.elsevier.com/locate/aca

# Identification of potential gene expression biomarkers for the surveillance of anabolic agents in bovine blood cells

Irmgard Riedmaier*, Ales Tichopad, Martina Reiter, Michael W. Pfaffl, Heinrich H.D. Meyer

*Physiology Weihenstephan, Technische Universitaet Muenchen, Weihenstephaner Berg 3, 85354 Freising, Germany*

### ARTICLE INFO

*Article history:*
Received 21 November 2008
Received in revised form 9 February 2009
Accepted 9 February 2009
Available online 20 February 2009

*Keywords:*
Anabolic agents
Trenbolone acetate
Estradiol
Biomarker
Gene expression
Quantitative real time reverse transcription polymerase chain reaction (qRT-PCR)
Principal components analysis

### ABSTRACT

In the EU, the use of anabolic steroids in food producing animals has been forbidden since 1988. The routine methods used in practice are based on the detection of hormonal residues. To overcome these routine methods, growth-promoting agents are sometimes administered at concentrations below the detection limit and new anabolic substances are designed. Therefore, new monitoring systems are needed to overcome the misuse of anabolic agents in meat production.

In this study, a new monitoring system was applied: the quantification of mRNA gene expression changes by quantitative real time reverse transcription polymerase chain reaction (qRT-PCR). Blood was selected as ideal tissue for biomarker screening. From the literature, it is known that steroid hormones affect mRNA gene expression of the different blood cells, which can easily be taken from the living animal.

In an animal trial, 18 Nguni heifers were separated to two groups of nine animals. One group served as untreated control and the other group was treated with a combination of trenbolone acetate plus estradiol for 39 days in order to allow the detection of the effect on mRNA expression in blood at three time points. Candidate genes used for developing a biomarker pattern were chosen by screening the actual literature for anabolic effects on blood cells.

It could be demonstrated that the combination of trenbolone acetate plus estradiol significantly influences mRNA expression of the steroid receptors (ER-α and GR-α), the apoptosis regulator Fas, the proinflammatory interleukins IL-1α, IL-1β and IL-6 and of MHCII, CK, MTPN, RBM5 and Actin-β. Advanced statistical analysis by Principal Components Analysis (PCA) indicated that these genes represent potential biomarkers for this hormone combination in whole blood.

© 2009 Elsevier B.V. All rights reserved.

## 1. Introduction

Growth-promoting agents like anabolic steroids or β-agonists are used in meat producing animals to improve weight gain and feed efficiency in order to increase the productivity and to reduce production costs [13,20]. Due to adverse effects of hormone residues for the consumer [4,27] the use of anabolic hormones for growth promotion is forbidden in the European Union since 1988 under Directive 88/146/EEC. Routine methods like immuno assays, either radio immuno assay (RIA) or enzyme immuno assay (EIA), and chromatographical methods combined with mass spectrometry are used to detect hormone residues [18,19,26,29]. To avoid detection of residues during routine control, growth-promoting agents are often administered in cocktails with such low amounts per agent that residues are below the detection limit [1]. Alternatively, new compounds, not yet included in testing programs, are used. Therefore, it is necessary to develop new monitoring systems to detect a broad range of agents at the lowest concentration that is used to get a growth-promoting effect. A potential way to develop a new monitoring system is to find gene expression biomarkers for the illegal use of anabolic steroids [23,24,28].

It is well known that steroid hormones influence biochemical pathways of different organs and tissues. mRNA expression of hormone dependent genes can be activated or suppressed.

Using appropriate specific and sensitive quantification methods, like quantitative real time reverse transcription polymerase chain reaction (qRT-PCR), such mRNA expression changes are measurable at very low levels. From the literature it is known that sex steroid hormones show physiological effects on the different blood cells [3,9,16].

The aim of this pilot study was to monitor the effects of a commercially available combination of trenbolone acetate plus estradiol on mRNA expression of selected target genes in bovine whole blood and to perform a bioinformatic evaluation in order to find potential biomarkers for the effective surveillance of this hormone combination.

---

* Corresponding author. Tel.: +49 8161 715552; fax: +49 8161 714204.
E-mail address: irmgard.riedmaier@wzw.tum.de (I. Riedmaier).

0003-2670/$ – see front matter © 2009 Elsevier B.V. All rights reserved.
doi:10.1016/j.aca.2009.02.014

## 2. Materials and methods

### 2.1. Animal experiment

18 healthy, nonpregnant, 2-year-old Nguni heifers were separated to two groups of nine animals each. One group was treated with Revalor H® (140 mg Trenbolone acetate plus 14 mg estradiol, Intervet, Isando, RSA) by implantation into the middle third of the pinna of the ear and one group was untreated serving as control.

Whole blood samples were taken at four time points. Predose samples were taken prior to treatment. Further samples were taken at day 2, day 16 and day 39 after treatment start. Blood samples (2.5 mL each) were transferred into PAXgene blood RNA tubes (BD, Heidelberg, Germany) gently shaken, incubated at room temperature for 2 h and stored at $-20\,°C$. At the same time points a complete blood count was done by the section of clinical pathology, University of Pretoria, South Africa, to control the health status of the animals. The animal attendance and blood sampling were done by the Onderstepoort Veterinary Institute (Onderstepoort, Pretoria, South Africa). The animals were housed and fed according to practice.

### 2.2. Total RNA extraction and quality determination

Total RNA from blood samples was extracted using the PAXgene Blood RNA Kit (Qiagen, Hilden, Germany) according to the manufacturer's instructions.

To quantify the amount of total RNA extracted, optical density ($OD_{260}$) was measured with the photometer (Eppendorf Biophotometer, Hamburg, Germany) for each sample. RNA purity was calculated with the $OD_{260/280}$ ratio.

RNA integrity and quality control was performed via capillary electrophoresis in the Bioanalyzer 2100 (Agilent Technology, Palo Alto, USA). Eukaryotic total RNA Nano Assay (Agilent Technology) was taken for sample analysis and the RNA Integrity Number (RIN) served as RNA quality parameter. Agilent Bioanalyzer 2100 calculated the RIN value based on a numbering system from 1 to 10 (1 being the most degraded profile, 10 being the most intact) for all samples.

### 2.3. RNA reverse transcription

Constant amounts of 1 μg total RNA were reverse transcribed to cDNA using the following master mix: 12 μL 5× Buffer (Promega, Mannheim, Germany), 3 μL Random Hexamer Primers (50 mM; Invitrogen), 3 μL dNTP Mix (10 mM; Fermentas, St Leon-Rot, Germany) and 200 U of MMLV H-Reverse Transcriptase (Promega) according to the manufacturer's instructions.

### 2.4. Specific primer design

All primers were designed using published bovine nucleic acid sequences of GenBank (http://www.ncbi.nlm.nih.gov/entrez/query.fcgi). Primer design and optimization was done with primer design program of MWG Biotech (MWG, Ebersberg, Germany) and primer3 (http://frodo.wi.mit.edu/cgi-bin/primer3/primer3_www.cgi) with regard to primer dimer and self-priming formation. Newly designed primers were ordered and synthesized at MWG Biotech. Primer testing was performed with three optional samples and a no template control (NTC contains only RNAse free water). To determine the optimal annealing temperature for each primer set a temperature gradient PCR was done. All used primers are listed in Table 1.

### 2.5. Quantitative PCR analysis

To analyze gene expression of candidate genes, qRT-PCR analysis was done using the iQ5 (Bio-Rad, Munich, Germany). Quantitative real-time RT-PCR was performed using MESA GREEN qPCR MasterMix Plus for SYBR® Assay w/fluorescein Kit (Eurogentec, Cologne, Germany) by a standard protocol, recommended by the manufacturer.

With the kit the master mix was prepared as follows: For one sample it is 7.5 μL MESA GREEN 2× PCR Master Mix, 1.5 μL forward primer (10 pmol $μL^{-1}$), 1.5 μL reverse primer (10 pmol $μL^{-1}$) and 3 μL RNase free water. For qPCR analysis 1.5 μL cDNA was added to 13.5 μL Master Mix. qPCR was performed in 96 Well Plates (Eppendorf, Hamburg, Germany) and pipetting was done by the epMotion 5075 (Eppendorf).

The following real-time PCR cycling protocol was employed for all investigated factors: denaturation for 5 min at 95 °C, 40 cycles of a two segmented amplification and quantification program (denaturation for 3 s at 95 °C, annealing for 10 s at primer specific annealing temperature listed in Table 1), a melting step by slow heating from 60 to 95 °C with a dwell time of 10 s and continuous fluorescence measurement. Threshold cycle (Ct) and melting curves were acquired by using the IQ5 Optical System software 2.0 (Bio-Rad). Only genes with clear melting curves were taken for further data analysis. Samples that showed irregular melting peaks were excluded from the quantification procedure.

### 2.6. Selection of candidate target genes

Candidate genes that might be biomarkers in blood were chosen by screening the respective literature for steroidal and inflammation related effects on blood cells.

It is known that steroid hormones affect the expression and mRNA stability of their receptors [10]. Therefore the steroid receptors androgen receptor (AR), estrogen receptor alpha (ER-α) and beta (ER-β) and the glucocorticoid receptor (GR-α) were chosen as candidate genes. It was already shown that testosterone influences the rate of apoptotic blood cells [3,9,16]. Different apoptosis regulators were included: TNF receptor superfamily member 6 (Fas), its ligand FasL, tumor necrosis factor receptor (TNFR) 1 and 2, their ligand tumor necrosis factor α (TNF-α), B-cell CLL/lymphoma 2 (BCL-2) and caspase 8 (Casp 8). Androgens are known to downregulate proliferation of lymphocytes [11,14]. Therefore a variety of pro- and anti-inflammatory interleukins (IL) (IL-1α, IL-1β, IL-6, IL-8, IL-10, IL-12B, IL-15) and the growth factors insulin-like growth factor 1 (IGF-1), tumor growth factor beta (TGF-β) and interferone gamma (IFN-γ) were analyzed. To determine if the treatment has as well an influence on the amount of the different white blood cells, the expression of the cell specific CD Antigens CD4 (T helper cells) and CD8 (cytotoxic T was measured. Further genes were the inflammatory factor nuclear factor of kappa light polypeptide gene enhancer in B-cells 1 (p105) (NFκB), major histocompatibility complex class II (MHC II), adrenergic beta kinase 2 (ADRBK2), actin-α 1 (ACTA1), creatin kinase (CK), jun oncogene (JUN), estrogen induced transcription factor (EITr), m yotropin (MTPN), tropomodulin 3 (TMOD3) and RNA binding protein 5 (RBM5). As reference gene candidates ubiquitin 3 (UB3), glyceraldehyde-3-phosphate dehydrogenase (GAPDH), actin-β (ACTB), histone and tyrosine 3-monooxygenase/tryptophan 5-monooxygenase activation protein, zeta polypeptide (YWHAZ) were measured whereas the UBC and GAPDH were chosen as best reference genes by using GenEx Ver 4.3.6 Software (MultiD Analyses AB, Gothenburg, Sweden).

### 2.7. Data analysis and statistics

Significant changes of the amount of the different blood cells between the two groups were determined using an unpaired t-test.

**Table 1**
List of primer pairs used for qRT-PCR analysis.

| Gene group | Gen | | Sequenz | Annealing temperature | Product length |
|---|---|---|---|---|---|
| Reference genes | GAPDH | for | GTC TTC ACT ACC ATG GAG AAG G | 60 °C | 197 bp |
| | | rev | TCA TGG ATG ACC TTG GCC CAG | | |
| | UB3 | for | AGA TCC AGG ATA AGG GAA GGC AT | 60 °C | 198 bp |
| | | rev | GCT CCA CCT CCA GGG TGA T | | |
| Steroid receptors | AR | for | CCT GGT TTT TCA ATG AGT ACC GCA TG | 62 °C | 172 bp |
| | | rev | TTG ATT TTT CAG CCC ATC CAC TGG A | | |
| | ERalpha | for | AGG GAA GCT CCT ATT TGC TCC | 60 °C | 233 bp |
| | | rev | GGT GGA TGT GGT CCT TCT C | | |
| | ERb | for | TTA GCC ATC CAT TGC CAG CC | 64 °C | 248 bp |
| | | rev | GCC TTA CAT CCT TCA CAC GAC | | |
| | GRa | for | TTC GAA GAA AAA ACT GCC CAG C | 64 °C | 190 bp |
| | | rev | CAG TGT TGG GGT GAG TTG TG | | |
| Apoptosis regulators | FasL | for | CAT CTT TGG AGA AGC AAA TAG | 60 °C | 205 bp |
| | | rev | GGA ATA CAC AAA ATA CAG CCC | | |
| | Fas | for | TGT TGT CAG CCT TGT CCT CC | 60 °C | 174 bp |
| | | rev | GTT CCA CTT CTA GCC CAT GTT C | | |
| | bcl-2 | for | ATG ACT TCT CTC GGC GCT AC | 60 °C | 245 bp |
| | | rev | CCG GTT CAG GTA CTC GGT CA | | |
| | TNFa | for | CCA CGT TGT AGC CGA CAT C | 60 °C | 155 bp |
| | | rev | CCC TGA AGA GGA CCT GTG AG | | |
| | TNFR1 | for | TCC AGT CCT GTC TCC ATT CC | 60 °C | 236 bp |
| | | rev | CTG GCT TCC CAC TTC TGA AC | | |
| | Casp 8 | for | TAG CAT AGC ACG GAA GCA GG | 60 °C | 294 bp |
| | | rev | GCC AGT GAA GTA AGA GGT CAG | | |
| | TNFR2 | for | AGCAGCACGGACAAGAGG | 60 °C | 220 bp |
| | | rev | CTGTGTCCCTCGTGGAG | | |
| Interleukins | IL-1a | for | CCT CTC TCT CAA TCA GAA GTC C | 64 °C | 142 bp |
| | | rev | CCA CCA TCA CCA CAT TCT CC | | |
| | IL-1β | for | TTC TCT CCA GCC AAC CTT CAT T | 60 °C | 198 bp |
| | | rev | ATC TGC AGC TGG ATG TTT CCA T | | |
| | IL-6 | for | GCT GAA TCT TCC AAA AAT GGA GG | 60 °C | 200 bp |
| | | rev | GCT TCA GGA TCT GGA TCA GTG | | |
| | IL-8 | for | ATG ACT TCC AAG CTG GCT GTT G | 60 °C | 149 bp |
| | | rev | TTG ATA AAT TTG GGG TGG AAA G | | |
| | IL-10 | for | TGA TGC CAC AGG CTG AGA ACC AC | 64 °C | 118 bp |
| | | rev | TCG CAG GGC AGA AAG CGA TGA C | | |
| | IL12B | for | TACACAGTGGAGTGTCAGGAG | 60 °C | 250 bp |
| | | rev | TCAGGGAGAAGTAGGAATGCG | | |
| | IL15 | for | TTCCATCCAGTGCTACTTGTG | 60 °C | 127 bp |
| | | rev | ACATACTGCCAGTTTGCTTCTG | | |
| CD antigens | CD 4 | for | TTC CTT CCC ACT CAC CTT CG | 63 °C | 132 bp |
| | | rev | ATC TTG TTC ACC TTC ACC TCT C | | |
| | CD 8 | for | AGA AGG TGG AGC TGC AAT GCG AG | 60 °C | 294 bp |
| | | rev | GCA AGA AGA CAG GCA CGA AGT TAC TGA AG | | |
| Growth factors | IGF-1 | for | CAT CCT CCT CGC ATC TCT TC | 62 °C | 238 bp |
| | | rev | CTC CAG CCT CCT CAG ATC AC | | |
| | TGFb | for | TTC ATG CCG TGA ATG GTG GCG | 60 °C | 167 bp |
| | | rev | ACG TCA CTG GAG TTG TGC GG | | |
| | IFNg | for | GCA GAT CCA GCG CAA AGC CAT AAA TG | 60 °C | 112 bp |
| | | rev | TCT CCG GCC TCG AAA GAG ATT CTG AC | | |
| Others | NFkB | for | GCC TGT CCT CTC TCA CCC CAT CTT TG | 60 °C | 149 bp |
| | | rev | ACA CCT CGA TGT CCT CTT TCT GCA CC | | |
| | YWHAZ | for | CAG GCT GAG CGA TAT GAT GAC | 60 °C | 141 bp |
| | | rev | GAC CCT CCA AGA TGA CCT AC | | |
| | ACTB | for | AAC TCC ATC ATG AAG TGT GAC | 60 °C | 202 bp |
| | | rev | GAT CCA CAT CTG CTG GAA GG | | |
| | Histon | for | ACTGCTACAAAAGCCGCTC | 62 °C | 233 bp |
| | | rev | ACTTGCCTCCTGCAAAGCAC | | |
| | ACTA1 | for | TAT TGT GCT CGA CTC CGG C | 63 °C | 160 bp |
| | | rev | GTC ACC AAG GAG TAG CCA C | | |
| | CK | for | ATG ACA GAC GAG GAG CAG CA | 60 °C | 183 bp |
| | | rev | ATG GAG ATG ACT CGG AGG TG | | |
| | ADRBK2 | for | ACC TAT GCC TTC CAC ACT CC | 60 °C | 121 bp |
| | | rev | CGT AAA ACC GCA TCT CCT TC | | |
| | MHC2 | for | AAC CTA CAG TGA CCA TCT CCC | 60 °C | 108 bp |
| | | rev | ACC ACC GAA CCT TGA TCT GG | | |
| | JUN | for | ATCAAGGCAGAGAGGAAGCG | 63 °C | 217 bp |
| | | rev | TTAGCATGAGTTGGCACCCG | | |
| | EfTr | for | GTTCCTCAATTCCGTCTTCATC | 60 °C | 216 bp |
| | | rev | TCACTGTTCTCCTCTCATCTC | | |
| | MTPN | for | ATT ATG CAG CAG ATT GTG GAC AG | 60 °C | 112 bp |

Table 1 (*Continued*)

| Gene group | Gen | | Sequenz | Annealing temperature | Product length |
|---|---|---|---|---|---|
| | | rev | TAGACGGCAGACAGAAGAGG | | |
| | TMOD3 | for | ATCTTGACCCTGAGAACGCC | 64 °C | 142 bp |
| | | rev | TCTTCCCTGTCCTTATGCTCC | | |
| | RBM5 | for | CCA TCA CGG AGA GCG ATA TTC | 60 °C | 164 bp |
| | | rev | TTTCTGATTGGCTTCCATCCAG | | |

Statistical description of the expression data as well as statistical tests were produced with Sigma Stat 3.0. The raw data were the Ct values obtained from each qPCR sample. Each qPCR sample was associated with a blood sample. Since the amplification efficiency was not known, the assumption of identical amplification efficiency 100% was made, allowing a more simple quantification model [15].

The Ct values of each gene were translated to normalized expression quantities using two reference genes in a form of normalization index. The normalization index was calculated as an arithmetic mean of the Ct values of the two reference genes:

$$\text{Reference index} = \text{mean}(Ct_{UBC}, Ct_{GAPDH}) \quad (1)$$

Then, an expression of every target gene was calculated relatively to the expression of the housekeeping gene as

$$\text{Normalized expression} = \frac{2^{\text{reference index}}}{2^{Ct\,\text{target gene}}} \quad (2)$$

where the 2 represents the 100% amplification efficiency. The normalized expressions of the time points 2, 16 and 90 days were then

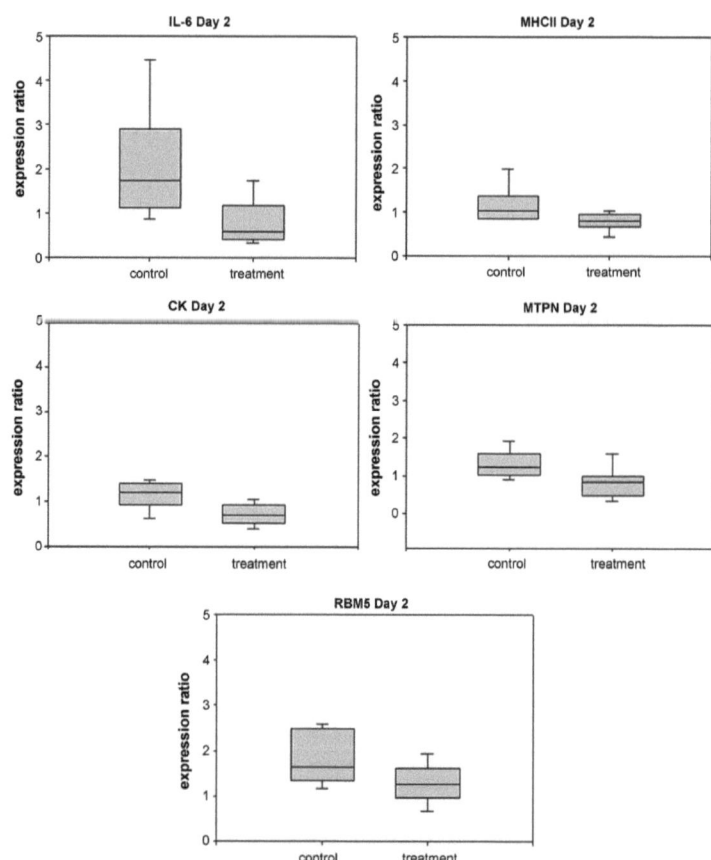

**Fig. 1.** Significant regulation for IL-6 (A), MHCII (B), CK (C), MTPN (D), and RBM5 (E) between control and treated samples after 2 days of treatment. Box plots show the group median and standard deviations.

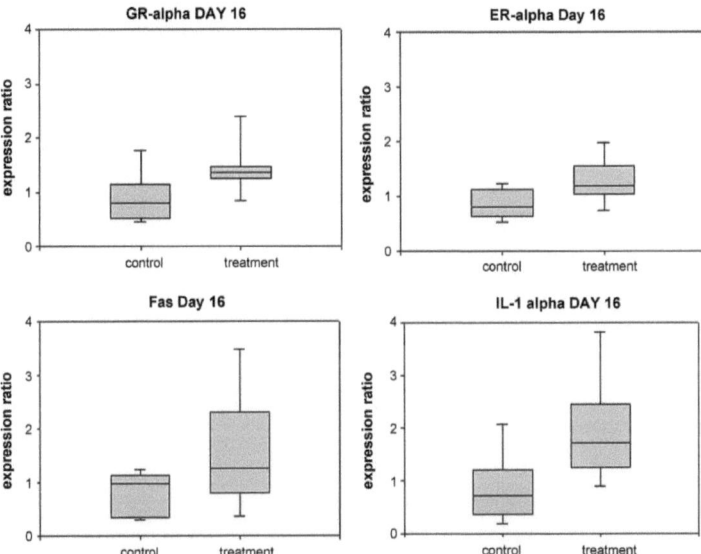

**Fig. 2.** Significant regulation for GR-α (A), ER-α (B), Fas (C) and IL-1α (D) between control and treated samples after 16 days of treatment. Box plots show the group median and standard deviations.

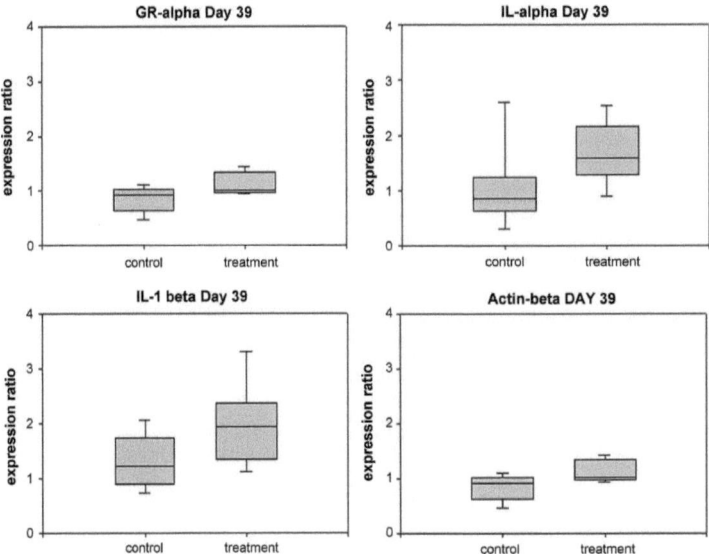

**Fig. 3.** Significant regulation for GR-α (A), IL-1α (B), IL-1β (C) and Actin-β (D) between control and treated samples after 39 days of treatment. Box plots show the group median and standard deviations.

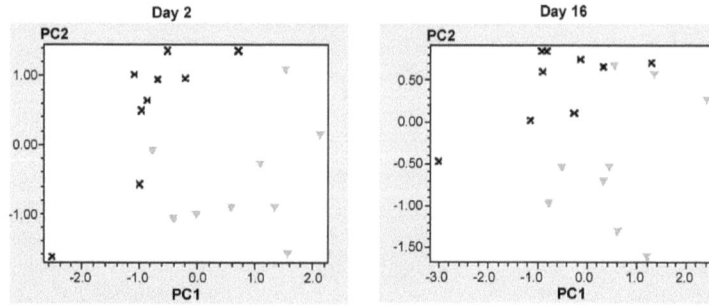
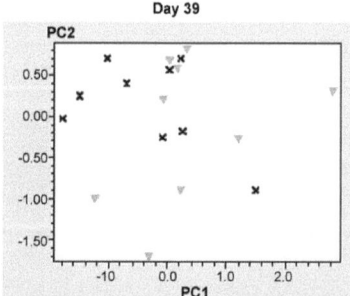

**Fig. 4.** Principal components analysis (PCA) for the eleven regulated genes GR-α, ER-α, Fas, IL-1α, IL-1β, IL-6, MHCII, CK, MTPN, RBM5 and Actin-β at the three different treatment time points. Animals of the control groups are represented by grey dots and animals of the treatment group are represented by black dots.

divided with the normalized expressions of the baseline (predose), generating the expression ratio $R$ as

$$R_{\text{timepoint/baseline}} = \frac{\text{Normalized expression}_{\text{timepoint}}}{\text{Normalized expression}_{\text{baseline}}} \quad (3)$$

The expression ratio $R$ was then analysed statistically using the $t$-test. The Box-whisker plot was constructed to facilitate visual screening of regulated genes (Figs. 1–3).

The objective of the statistical analysis was to disclose genes with significant regulation between control group and treatment group. Hence, this study is to be considered as purely explorative whereas significant findings here indicate candidate biomarkers.

To disclose multivariate response to the treatment, the method of principal component analysis (PCA) was employed using GenEx v. 4.3.6 (MultiD Analyses AB). PCA involves a mathematical procedure that transforms a number of variables (here normalized expression values) into a smaller number of uncorrelated variables called principal components. By this the dimensionality of the data is reduced to a number of dimensions that can be plotted in a scatter plot, here two dimensions. The first principal component accounts for as much of the variability in the data as possible, and each succeeding component accounts for as much of the remaining variability as possible. Normalized expression values of all responding genes were taken as the initial variables and reduced to two principal components only, facilitating thus resolution of treatment clusters in the scatter plot (Fig. 4) [12].

## 3. Results and discussion

### 3.1. RNA integrity

Good RNA quality is important for the overall success of RNA based analysis methods like real time qRT-PCR [7,8,21,25]. The RNA degradation level was determined using the lab-on-a-chip technology of the Agilent Bioanalyzer 2100 (Agilent Technologies). The mean (±std. dev.) RIN value of the blood samples was 8.3 ± 0.3 indicating fully integer total RNA.

### 3.2. Primer testing and gel electrophoresis

Primer pairs of 38 genes were successfully used in quantitative RT-PCR analysis to get single peaks and uniform melting curves.

**Table 2**
List of $p$ values for the regulation of the amount of the different blood cells.

| Time point | White blood cell count | Lymphocytes | Monocytes | Eosinophils | Basophils |
|---|---|---|---|---|---|
| Predose | 0.5347 | 0.9263 | 0.1273 | 0.1914 | 0.1691 |
| Day 2 | 0.2827 | 0.8051 | 0.8979 | 0.3663 | – |
| Day 16 | 0.9310 | 0.7601 | 0.0848 | 0.3551 | 0.3927 |
| Day 39 | 0.3758 | 0.5106 | 0.4026 | 0.0690 | 0.8353 |

**Table 3**
Significant mRNA expression changes. p values and x-fold regulation between steroid treatment and control group.

| Gene group | Gene | Time point | p value | Fold regulation |
|---|---|---|---|---|
| Steroid receptors | GR-a | Day 16 | 0.0159 | 1.597 |
|  | GR-a | Day 39 | 0.0273 | 1.345 |
|  | ER-a | Day 16 | 0.0106 | 1.509 |
| Apoptosis regulators | Fas | Day 16 | 0.0463 | 1.978 |
| Interleukins | IL-1a | Day 16 | 0.0108 | 2.268 |
|  | IL-1a | Day 39 | 0.0364 | 1.650 |
|  | IL-1b | Day 39 | 0.0412 | 1.475 |
|  | IL-6 | Day 2 | 0.0125 | 0.434 |
| Others | MHCII | Day 2 | 0.0219 | 0.682 |
|  | CK | Day 2 | 0.0046 | 0.637 |
|  | MTPN | Day 2 | 0.0129 | 0.621 |
|  | RBM5 | Day 2 | 0.0353 | 0.637 |
|  | Actin-b | Day 39 | 0.0095 | 1.345 |

### 3.3. Haemogram

The haemograms indicate that the animals were healthy. The white blood cell count and the amount of lymphocytes, monocytes, eosinophil, and basophil granolucytes ranged in physiological levels with no significant changes between both treatment groups (p values are listed in Table 2). Therefore significant changes in mRNA expression can be interpreted as real changes in gene expression and are not due to changes in the blood cell, especially the mRNA expressing white blood cells.

### 3.4. qRT-PCR results and data analysis

Significant regulation of gene expression of the treatment group compared to the control group could be identified for IL-6, MHC II, CK, MTPN and RBM5 after 2 days (Fig. 1), for GR-$\alpha$, ER-$\alpha$, Fas and IL-1$\alpha$ after 16 days (Fig. 2) and for Actin-$\beta$, GR-$\alpha$, IL-1$\alpha$ and IL-1$\beta$ after 39 days of treatment (Fig. 3). The resulting p values and the regulation ratio between control and treatment are listed in Table 3.

In the box-whisker plots it can be observed that there are also differences of gene expression in the control group compared to baseline. This reflects the natural variability of the non-induced expression in each studied subject.

The number of quantified genes was yet too less to draw conclusions on the different pathways but anyhow first physiological declarations can be made and genes that could act as potential biomarkers could be identified.

The steroid receptors GR-$\alpha$ and ER-$\alpha$ show an up-regulation in the treatment group compared to the control. GR-$\alpha$ is up-regulated at day 16 and day 39 whereas ER-$\alpha$ is only up-regulated at day 16. Trenbolone acetate has an antiglucocorticoid effect via binding to the glucocorticoid receptor [2,17,22]. It is already shown that anabolic steroids influence the mRNA expression of GR-$\alpha$ and ER-$\alpha$ in muscle tissue [24]. The applied hormone combination acts via both regulated steroid receptors. The up-regulation of both receptors indicate that in white blood cells the expression of these receptors is stimulated by its ligands.

The interleukins IL-1$\alpha$ and IL-1$\beta$ are up-regulated. IL-1$\alpha$ is up-regulated at day 16 and day 39 whereas IL-1$\beta$ is only regulated after 39 days of treatment. IL-1$\alpha$ and IL-1$\beta$ are produced by macrophages, monocytes and dendritic cells. During infection they induce the release of other cytokines. The expression of IL-1$\beta$ can be induced by IL-1$\alpha$. This could be an explanation why IL-1$\alpha$ is up-regulated after 16 days of treatment whereas IL-1$\beta$ is only up-regulated after 39 days of treatment [5,6].

Principal components analysis (PCA) is a technique used to reduce multidimensional data sets to lower dimensions for analysis. This statistical method was used to determine whether there is a clustering between control and treatment group. Fig. 4 was obtained by plotting all samples of the two groups in the different time points by their two principal components obtained from the 11 regulated genes. Each group was marked by a color. Black crosses represent samples of the control group and grey triangles show the samples of the treatment group. At days 2 and 16 of treatment it can be observed that both group arrange together and that a difference between control and treatment group can be monitored.

This observation is a first hint that it is possible to get a gene expression pattern opening the possibility to develop a screening method to control the misuse of anabolic hormones in cattle via blood cells. It will be a question of further in vivo trials to determine, if the suggested parameters are independent of breed, nutrition, age, gender and immune status of the animals, and whether they are sensitive enough to uncover low dosages.

## 4. Conclusions

This pilot study demonstrates that gene expression analysis could be a promising complement to hormone residue analysis for surveillance of hormone misuse in animal production. It could be shown that the combination of trenbolone acetate plus estradiol influences gene expression of 11 genes out of 38 tested candidate genes. Using principle component analysis such regulated genes could act as first biomarkers to discover the illegal use of anabolic hormones in cattle.

### Acknowledgements

We thank the Onderstepoort Veterinary Institute, Pretoria, Republic South Africa, for supporting this study. Special thanks go to Azel Swemmer and Kobus van der Merwe for study performance.

### References

[1] A.M. Andersson, N.E. Skakkebaek, Eur. J. Endocrinol. 140 (6) (1999) 477.
[2] E.R. Bauer, A. Daxenberger, T. Petri, H. Sauerwein, H.H. Meyer, APMIS 108 (12) (2000) 838.
[3] M. Cutolo, S. Capellino, P. Montagna, P. Ghiorzo, A. Sulli, B. Villaggio, Arthritis Res. Therapy 7 (5) (2005) R1124.
[4] A. Daxenberger, D. Ibarreta, H.H.D. Meyer, Hum. Reprod. Update 7 (3) (2001) 340.
[5] C.A. Dinarello, Eur. Cytokine Netw. 5 (6) (1994) 517.
[6] C.A. Dinarello, FASEB J. 8 (15) (1994) 1314.
[7] S. Fleige, M.W. Pfaffl, Mol. Aspects Med. 27 (2–3) (2006) 126.
[8] S. Fleige, V. Walf, S. Huch, C. Prgomet, J. Sehm, M.W. Pfaffl, Biotechnol. Lett. 28 (19) (2006) 1601.
[9] S.A. Huber, J. Kupperman, M.K. Newell, Lupus 8 (5) (1999) 384.
[10] N.H. Ing, Biol. Reprod. 72 (6) (2005) 1290.
[11] J.D. Jacobson, M.A. Ansari, Endocrinology 145 (1) (2004) 330.
[12] M. Kubista, J.M. Andrade, M. Bengtsson, A. Forootan, J. Jonak, K. Lind, R. Sindelka, R. Sjoback, B. Sjogreen, L. Strombom, A. Stahlberg, N. Zoric, Mol. Aspects Med. 27 (2–3) (2006) 95.

[13] I.G. Lange, A. Daxenberger, H.H. Meyer, APMIS 109 (1) (2001) 53.
[14] D. Lehmann, K. Siebold, L.R. Emmons, H. Muller, Clin. Immunol. Immunopathol. 46 (1) (1988) 122.
[15] K.J. Livak, T.D. Schmittgen, Methods 25 (4) (2001) 402.
[16] R.W. McMurray, S. Suwannaroj, K. Ndebele, J.K. Jenkins, Pathobiology 69 (1) (2001) 44.
[17] H.H.D. Meyer, APMIS 109 (1) (2001) 1.
[18] H.H.D. Meyer, S. Hoffmann, Food Addit. Contam. 4 (2) (1987) 149.
[19] H.H.D. Meyer, L. Rinke, I. Dursch, J. Chromatogr. 564 (2) (1991) 551.
[20] C. Moran, J.F. Quirke, D.J. Prendiville, S. Bourke, J.F. Roche, J. Anim. Sci. 69 (11) (1991) 4249.
[21] M.W. Pfaffl, S. Fleige, I. Riedmaier, Biotechnol. Biotechnol. EQ. 22 (2008) 829.
[22] J. Pottier, C. Cousty, R.J. Heitzman, I.P. Reynolds, Xenobiotica 11 (7) (1981) 489.
[23] M. Reiter, M.W. Pfaffl, M. Schoenfelder, H.H.D. Meyer, Biomarker Insights 4 (2009) 1.
[24] M. Reiter, V.M. Walf, A. Christians, M.W. Pfaffl, H.H. Meyer, Anal. Chim. Acta 586 (1-2) (2007) 73.
[25] A. Schroeder, O. Mueller, S. Stocker, R. Salowsky, M. Leiber, M. Gassmann, S. Lightfoot, W. Menzel, M. Granzow, T. Ragg, BMC. Mol. Biol. 7 (2006) 3.
[26] M.L. Scippo, G. Degand, A. Duyckaerts, G. Maghuin-Rogister, P. Delahaut, Analyst 119 (12) (1994) 2639.
[27] S.H. Swan, F. Liu, J.W. Overstreet, C. Brazil, N.E. Skakkebaek, Hum. Reprod. 22 (6) (2007) 1497.
[28] L. Toffolatti, G.L. Rosa, T. Patarnello, C. Romualdi, R. Merlanti, C. Montesissa, L. Poppi, M. Castagnaro, L. Bargelloni, Domest. Anim. Endocrinol. 30 (1) (2006) 38.
[29] L.A. van Ginkel, J. Chromatogr. 564 (2) (1991) 363.

Die VDM Verlagsservicegesellschaft sucht für wissenschaftliche Verlage abgeschlossene und herausragende

## Dissertationen, Habilitationen, Diplomarbeiten, Master Theses, Magisterarbeiten usw.

für die kostenlose Publikation als Fachbuch.

Sie verfügen über eine Arbeit, die hohen inhaltlichen und formalen Ansprüchen genügt, und haben Interesse an einer honorarvergüteten Publikation?

Dann senden Sie bitte erste Informationen über sich und Ihre Arbeit per Email an *info@vdm-vsg.de*.

**Sie erhalten kurzfristig unser Feedback!**

VDM Verlagsservicegesellschaft mbH
Dudweiler Landstr. 99  Telefon  +49 681 3720 174
D - 66123 Saarbrücken  Fax  +49 681 3720 1749
**www.vdm-vsg.de**

Die VDM Verlagsservicegesellschaft mbH vertritt

Printed by Books on Demand GmbH, Norderstedt / Germany